ation

MW00782947

GAMER NATION

Video Games and American Culture

JOHN WILLS

Johns Hopkins University Press BALTIMORE

© 2019 Johns Hopkins University Press
All rights reserved. Published 2019
Printed in the United States of America on acid-free paper
9 8 7 6 5 4 3 2 1

Johns Hopkins University Press
2715 North Charles Street
Baltimore, Maryland 21218-4363
www.press.jhu.edu

Library of Congress cataloging data is available.

ISBN-13: 978-1-4214-2870-3 (paperback: alk. paper)
ISBN-10: 1-4214-2870-9 (paperback: alk. paper)
ISBN-13: 978-1-4214-2869-7 (electronic)
ISBN-10: 1-4214-2869-5 (electronic)

A catalog record for this book is available from the British Library.

*Special discounts are available for bulk purchases of this book. For more information,
please contact Special Sales at 410-516-6936 or specialsales@press.jhu.edu.*

Johns Hopkins University Press uses environmentally friendly book materials, including
recycled text paper that is composed of at least 30 percent post-consumer waste, whenever
possible.

CONTENTS

ACKNOWLEDGMENTS

THE PROCESS of researching and writing this book has involved exploring all kinds of American media and materials. Alongside more traditional sources such as newspapers and historical archives, online discussion groups, fansites, along with the extended play of video games themselves, have all proved vital to this project. A number of game designers gave their time to answer questions on specific video game titles. I thank Nolan Bushnell, founder of Atari, Don Rawitsch, creator of *The Oregon Trail* (1971), Jim Storer, programmer of *Lunar Lander* (1969), Scott Harrison, the graphics artist for *Tapper* (1983), as well as Simon Taylor, who worked on the title *New York Blitz* (1982), for all graciously taking the time to reply to my questions. For materials on the historic Oregon Trail and George Armstrong Custer, I thank Marva Felchin, director of Libraries and Archives at the Autry Museum of the American West in Los Angeles, as well as staff at the Huntington Library in San Marino.

Chapter Two, exploring depictions of the "Wild West" in video games, draws from my article "Pixel Cowboys and Silicon Goldmines," published in 2008 for the *Pacific Historical Review*. I thank editor Marc Rodriguez for permission to use sections from the original article. The British Academy, London, funded research linked with this project, and gave me the opportunity to exhibit my work at the Academy's first annual Summer Showcase in June 2018, whereby, across three days, guests chatted with me about the cultural impact of video games and also played a range of vintage amusement machines, with a 1966 Williams pinball machine and two Atari VCSs proving most popular. My dear thanks to event organizers Johanna Empson and Marisa Smith and the British Academy team.

At Johns Hopkins University Press, I thank my editor Robert J. Brugger, who first showed interest in the project and was patient enough to wait for it to come to fruition, along with Matt McAdam, for seeing the project through, and editorial assistant William Krause. At the University of Kent, I thank my good friend and visual scholar Tom Lawrence for countless conversations about gaming (and some wild Pokémon hunting sessions across the summer of 2017), as well as his valuable input into the manuscript. Colleague, American scholar, and fellow gamer Nicholas Blower provided original insights into video game culture, especially the mysterious cults found in *Grand Theft Auto V* (2013), and commented on individual chapters. My research assistant Hollie Bramwell assisted with a range of tasks across 2017 and 2018, and both Nic and Hollie helped out at the British Academy Showcase. Game Studies scholar Esther Wright at Warwick University also took time out to read the final draft of the manuscript.

The book draws on a number of research papers given at annual conferences hosted by the Native Studies Research Network (NSRN), British Association for American Studies (BAAS), and Historians of the Twentieth Century United States (HOTCUS), along with symposiums at Christ Church University, the University of Sussex and the University of East Anglia. These events brought me into contact with some superb Game Studies scholars, including Ewan Kirkland, Alan Meades, Chris Pallant, Adam Chapman, and Emily Marlow.

My partner, Samantha Robinson, and my toddler son, Jude, coped patiently with a project that seemed to morph between typing and gaming all too fluidly.

This work is dedicated to my mother and late father, who introduced me to video games. One of my first memories is playing a cowboy-themed shooting title at an amusement arcade during a family trip in the late 1970s. Despite its significant cost, my parents bought me a Commodore Vic 20, my first computer, back in 1982. Although I was meant to use the machine for education, it soon became more of a plaything. I remember playing *New York Blitz* on the system, along with *Ski* (1983), *JetPac* (1983), and android-invasion title *Amok!* (1981). I became friends with Tim King and Chris Lewis at college around that time, and we played a huge range of computer games at each other's houses, from space adventure titles *Elite* (1984) and *Laser Squad* (1988) to fighting games *The Way of the Exploding Fist* (1985) and *IK+* (1987). Those early experiences led to an interest in gaming that endures today. Some of my favorite memories over the past four decades involve playing computer and video games with people. But it all began with a classic Wild West scene and some pixel gunslingers.

Gamer Nation

Introduction: A New Realm of Play

I N 1975, design engineer Dave Nutting completed work on a new arcade machine for Midway, the Chicago-based amusement firm. A version of Taito's *Western Gun*, a recent Japanese arcade machine, Nutting's *Gun Fight* depicted a classic showdown between gunfighters. Two stick-like pixel figures roamed a sparse digital landscape, with only sagebrush rolling vertically down the screen. Rich in Western folklore, the game seemed perfect for the US market. Midway shipped the first machines to a variety of amusement arcades across the country. The old frontier artwork of the tall game cabinets stood out in the neon-lit arcades, as if transported there from a different age. At the same time, a Wild West duel was something that every American could instantly recognize and relate to. Players easily picked up the format and became pistol-wielding pixel cowboys.

Gun Fight sold more than 8,000 units.[1] One of the first successful arcade titles, Midway's machine helped introduce a nation to video gaming. In the early 1970s, few people played arcade games, owned a game console, or had access to a computer. The Magnavox Odyssey (1972), the first game console for the home, was sold only through Magnavox television dealers and had restricted returns. Computers remained prohibitively expensive and thus were mostly found in offices. Scattered across arcades and bars, video game machines were unfamiliar and exotic. Apart from Atari's *Pong* (1972), a well-received table tennis game, video games were few and far between. With very little instruction, however, *Gun Fight* showed Americans how to become gamers in the new realm of digital play.

Midway's *Gun Fight* contributed to a growing fascination with video games that marked the late 1970s and early 1980s. First arcade machines, then home consoles and microcomputers, vied for popular attention. A new entertainment media, video games featured tantalizing worlds to explore and quickly transformed people into gamers. Teenagers fed coins into arcade machines at the mall for fast-paced five-minute sessions on action-based titles, tackling alien invasion head on in *Space Invaders* (1978) or avoiding ghosts in the maze game *Pac-Man* (1980). Families played rudimentary sports, driving, and combat games at home on their television with Mattel's Intellivision and Atari VCS consoles. Milton Bradley's Vectrex home machine featured a black-and-white screen, color overlays, and an arcade-like control pad with big red buttons and a joystick. Schoolkids carried portable electronic games with them, such as Mattel's *Auto Race* (1976), a simple driving game with a bright LED denoting the car, and Nintendo's Game & Watch system, which served as a clock, alarm, and miniature game, all on a pocket-sized LCD-based screen. Promising a wide range of domestic uses, from managing home accounts to organizing chores, 8-bit computers such as the Apple II and Commodore 64 owed their popularity in part to their expansive catalog of games. With the games often on cassette or floppy disk, home users waited patiently for titles to "load" and for the fun to begin. Video games infiltrated all kinds of environments, from the shopping mall to the school, from the office to the bedroom.

Early video games promised to transport their audiences to new realms of entertainment courtesy of microprocessors. Games machines seemed futuristic and revolutionary; they peered into America's tomorrow. Markers of an imminent high-tech age, they seemed part of a greater story of technological advancement. Nolan Bushnell's *Computer Space* (1971), a space combat game enclosed in a stylish fiberglass cabinet, was featured in the science-fiction movie *Soylent Green* (1973), a film set in 2022 New York City. An early example of product placement, the machine took pride of place in the living room of leading New York businessman William R. Simonson (Joseph Cotten). In one scene, concubine Shirl (Leigh Taylor-Young) expressed her excitement over shooting down five saucers with a rocket, purring "Thank you for the toy" to Simonson (who passes on playing). The film situated video games as desirable yet frivolous commodities for future elites, the playthings of playthings.

Video games also drew on familiar activities and settings for their content. Titles provided pixelated versions of all manner of US hobbies and sports, from billiards to baseball. They told decidedly conventional American stories. Midway's *Gun Fight* owed much of its success to its harnessing of established Wild West folklore. Having played Cowboys and Indians at school and watched *Gunsmoke* (1955–75) on television, most Americans already knew the popular story of the United States' wild frontier. The transition from watching a cowboy on television to controlling a gunfighter on screen seemed almost evolutionary. The myth of the West (and its appealing explanation of American nationhood) helped sell early video games and fueled the mass fascination with digital play. At the same time, *Gun Fight* asserted its own distinctive narrative of the Western endeavor: a frontier without Native Americans or settlers, defined by a hail of bullets. The game told its own story about America. Players watched, listened, and returned fire.

A Nation of Gamers

In 2004, while reviewing a book on US sports, a journalist for the *New York Times* declared that the National Football League represented "the real national pastime."[2] America *is* a Gamer Nation, but not because of games of the cowhide or vulcanized rubber variety. The real national pastime involves joypads and keyboards, not shoulder pads and footballs. America is a nation of video gamers. In 2016, around seventeen million viewers watched regular NFL games on television, and around one million teenagers played competitively on football teams in high school.[3] By contrast, an estimated 185 million people played video games. In a December 2015 survey Gaming and Gamers, Pew researchers found that half of American adults and a staggering 97 percent of teenagers played video games. Around five million "serious gamers" expended more than forty hours a week on titles such as *The Elder Scrolls V: Skyrim* (2011) and *World of Warcraft* (2004+).[4] The *Washington Post* has reported a growing phenomenon of young adult men who find more reward in playing video games than developing their careers, some eschewing the job market entirely (what might be called "play at home" men).[5] Four out of five households owned a games console, while "casual gaming," associated with smartphone titles such as *Angry Birds* (2009+), proved even more widespread.[6]

The cultural impact of the "new media" of video games is immense. As Simon Egenfeldt-Nielsen et al. contend, "In the historical blink of an eye, video games have colonized our minds and invaded our screens."[7] Over the past four decades, US companies Atari, Microsoft, and Electronic Arts, along with Japanese companies Nintendo, Sega, and Sony, have forged a new world of consumer entertainment. The US video game industry hovers at around $25 billion in worth. In the entertainment industry, video games compete with film and literature in popularity. Video games rival Hollywood movies in production values, promotional campaigns, and profit margins. By 2018, Rockstar's *Grand Theft Auto V* (2013), a mature-rated game about criminal life in a fictional Los Angeles, had sold ninety million units and accumulated $6 billion in sales, making it the most successful entertainment product ever released, substantially ahead of Victor Fleming's *Gone with the Wind* (1939) and George Lucas's *Star Wars* (1977). According to the Davie-Brown Index for brand recognition, the Pac-Man gaming icon rated at 94 percent consumer recognition in 2008, putting the dot-eating yellow character (originally named Puck Man, owing to his likeness to a hockey puck) ahead of a range of leading Hollywood film stars.[8]

Consecutive phases of gaming have contributed to distinct subcultures and social groups, from arcade culture, hackers, and teenage mall culture to online player communities in the 2010s. Video games are connected to realms traditionally not associated with play, including politics, health, education, and military warfare. Introduced in 2002, the US Army–endorsed first-person shooter game *America's Army* functioned as an interactive propaganda piece for the military, encouraging US citizens to enroll in the armed services. In both the 2012 and 2016 elections, lobbying groups employed mobile application (app) games to persuade voters. Despite their "armchair" reputation, games have encouraged Americans to become active, from the home-fitness programs in Nintendo's *Wii Fit* (2008) to the outdoor play in *Pokémon Go* (2016). In kindergarten, titles such as *Barbie Dress-up* teach toddlers motor skills, while in high school, educational titles teach history. Senior citizens can keep their mental skills honed with Nintendo's *Brain Age* (2005+), games inspired by the work of Japanese neuroscientist Ryuta Kawashima. From generation to generation, from birth to retirement, gaming touches most aspects of American life.

But where did this Gamer Nation come from? The advent of electronic gaming and the emergence of a Gamer Nation owes much to broader technological, corporate, and cultural forces in the late twentieth century. The popularity of video games stems from new technologies, US inventiveness, corporate growth, consumer trends, and a national shift toward recreation and play. The early video game industry emerged from electro-mechanical entertainment found at amusement parks and theme parks, including Coney Island and Disneyland, with many pinball manufacturers embracing the move to producing and marketing arcade machines by the late 1970s. The television screen directly inspired gaming, with early interfaces coupling black-and-white screens with rudimentary joystick devices, as well as the term "video game" (in the 1970s sometimes called "tele-game"), a combination of view screen (video) and play (game). Breakthroughs at military installations, university campuses, and technology companies regarding chip technology, programming methods, and machine connectivity allowed the development of games of increasing complexity. The rise of gaming also reflected the computer revolution, the emergence of Silicon Valley, and the dawn of a digital age. Both Steve Jobs and Steve Wozniak worked on the Atari arcade machine *Breakout* (1976), a game about knocking down a wall of bricks, before forming Apple Computers. As Michael Newman notes, "During their emergence, video games were objects without fixed meanings," and their identity in the 1970s fused with ideas about all manner of technologies.[9]

Video games represented one component in a broad, fast-emerging circuitry of digital entertainment in the late twentieth century. Game machines joined digital watches, pocket calculators, and early home computers as the United States went electronic. The US consumer largely embraced a raft of new media and digital technology. The rise of video games reflected a thriving digital culture industry and its expanding product line. Sometimes game content openly displayed corporate investment and synergy at work. Budweiser initially sponsored bar-based game *Tapper* (1983); Johnson & Johnson produced the "advergame" *Tooth Protectors* (1983), which promoted good dental hygiene; and Coca-Cola created its own version of the arcade hit *Space Invaders* (1978) named *Pepsi Invaders* for an early 1980s sales convention. Video game players appeared caught in a world of consumer electronics. The "video game turn" reflected the increasing assumption that digital consumer products could fulfill all kinds of psychological needs.

In the 1930s, cultural theorist Johan Huizinga defined *Homo ludens* as man the player, an archetype based around the native trickster character.[10] *Homo ludens* have also turned out to be man the gamer. Video games became ever more popular in the late twentieth century because of an increasing interest in recreation more generally. In many ways, the twentieth century as a whole might be seen as the "ludic century," a time in which play became a normalized part of everyday American life. Aided by a growing middle class, greater affluence, and a significant increase in leisure time, the average American in the post-1945 period welcomed recreational opportunities on an unbridled scale. Bowling and board games had their heydays in the 1950s, while conventional sports such as baseball began being televised regularly. In the 1970s, video games built on this recreational impetus. They updated familiar games, puzzles, and sports, as well as digitizing war games. They also reflected a broader gamification of America that continues into the present day.[11]

America as a Game

Along with the structural forces that helped bring video games into mainstream culture, the success of these games was a result of them regularly co-opting American symbols and stories. As in the case of *Gun Fight*, the rise of a Gamer Nation reflected the ability of video games to speak to American myth, culture, and society. When Nolan Bushnell introduced the first commercial arcade machine, *Computer Space*, at the Music Operators of America (MOA) Music and Amusement Machines Expo in October 1971, businessmen scoffed at the possibility of the machine competing with, let alone replacing, the established arcade favorite, pinball. The trade audience found Bushnell's machine confusing and expensive, and as Bushnell remembers, "it was poorly received." On release, *Computer Space* struggled for business in the traditional commercial territory of the American bar, deemed too tricky to understand and a distraction from banter and beer. Only 1,500 units sold. *Computer Space* proved too abstract in concept and too complicated for its players. Customers struggled to comprehend the novel technology and rejected the unfamiliar narrative. Bushnell later reflected, "Sometimes you can do too many innovations at the same time . . . the complexity of the game limited its market."[12] *Computer Space* failed to speak to everyday America and instead seemed something straight out of the future.

Bushnell followed *Computer Space* with a very different type of game, *Pong* (1972). This new arcade title dropped the complex interface and simplified the game content: it recreated table tennis by putting two bats and a ball on screen. *Pong* was a game that every American could easily understand and even master. In 1975, Bushnell's company, Atari, released a home version through the Sears catalogue. "Pong is one of the strongest items we have this Christmas," reported a Sears representative that December. *Pong* became the biggest video game hit of the decade. Other companies copied Bushnell's concept, with a rush of bat-and-ball-type electronic games released onto the market.[13]

During the late 1970s, the American public could choose from a growing range of games at arcades and at home. Titles replicated US history, hobbies, and sports, as well as tackling larger hopes and fears within society. Titles about America helped Americans understand video games while they negotiated the joystick controls, distinctive pixel imagery, and often numerical goals. In addition to immersive game worlds drawn from pure fantasy and abstraction, video games took their cues from the real world of the Cold War and contemporary US life. In albeit limited ways (and also basic graphics), video games began to simulate, and thus comment on, all kinds of American themes and topics. As well as fighting deadly duels in *Gun Fight* (1975) and its follow-up, *Boot Hill* (1977), players could become bartenders serving beverages in *Tapper* (1983), play on a gridiron with Mattel's portable title *Football* (1977), and cycle around suburbia delivering newspapers in *Paperboy* (1985). Games mimicked and, in turn, toyed with the American experience. They also developed new stories about the nation. The game *Deus Ex Machina* (1984), noted for its unconventional perspective from the point of a view of a "defect" in a machine, dared to imagine a cyber-world and to consider what androids think about.[14] Video games sold imagined pasts and futures, filled with avatar bodies and digital homesteads. They sold not only a next frontier, but also a new fantasy of America.

The same year that *Gun Fight* entered arcades, *Death Race 2000* opened at cinemas. It was an ultra-violent dystopian movie with cult pretensions produced by Roger Corman and directed by Paul Bartel.

The movie poster promoted the film based on its shock value, announcing, "In the year 2000, hit and run driving is no longer a felony. It's the national sport."[15] In 1976, Californian company Exidy released an arcade machine, *Death Race*, likely inspired by the movie. Designed by Howell Ivy, the game instructed players to target fleeing gremlins on the road and drive over them at speed. The stick-like gremlins resembled people, and the working title of the game was *Pedestrian*. The game conceptualized the American highway as a veritable kill zone. It depicted future American society as violent and shocking, with the player an active participant in a dystopian digital world. Games now entertained what could go wrong with the country and shared a disturbing story.

Death Race received a slew of complaints over its lurid depiction of violence. The magazine *Family Safety*, produced by the National Safety Council, deemed the title "sick, sick, sick," while behavioral psychologist Gerald Driessen explained, "On TV, violence is passive. In this game, a player takes the first step to creating violence."[16] The notion of video games as facilitators of deviant behavior, especially violence, took hold. The idea of video games as deviant was aided by the locations in which they were played. In the mid-1970s, gamers congregated in often dark and dingy arcades, urban geographies situated outside mainstream life. People did not want a Death Race America, nor did they want players acting it out in dark spaces in their hometowns. As a result, video games initially operated on the fringes of society.

Alongside their associations with family fun and exciting new technology, video games became identified with notions of American decline and decay. The *Death Race* controversy fed growing suspicions over the negative impacts of gaming. In the early 1980s, Ronnie Lamb from the Long Island Parent-Teacher Association led a well-publicized campaign against emerging digital play. Charging that "these games are corrupting our youth," Lamb linked video game culture to teenage antisocial behavior, including littering, crime, gambling, aggression, and "unacceptable language." Lamb targeted arcade owners, calling the people who "open these places . . . the absolute dregs, and they are battling motherhood and apple pie." According to Lamb, video game America posed a fundamental threat to American values. Vowing to close down the gaming business, Lamb proclaimed, "We will be victorious."[17]

Many American arcades closed in the 1980s, but due to a loss of business, not to citizen protest. A crash in the domestic market in 1983, the advent of affordable home computing, and the availability of new home consoles by Japanese manufacturers Nintendo and Sega all contributed to the decline of the American arcade. At the same time, the video game industry continued to build momentum over the decades that followed, ultimately competing alongside television and film as the nation's primary entertainment. Media giants Sony and Microsoft launched their own game consoles, PlayStation (1994) and Xbox (2001), respectively, that sold millions of units, while Nintendo's Game Boy (1989) perfected mobile gaming. Home computers remained popular for complex strategy and role-playing titles. In the 2000s, smart phones broadened the market for video games. The gaming business has grown exponentially and, with it, the number of digital stories. Lamb proved far from victorious in her crusade.

Deciphering the Code

Corporate sales targets, the physical spaces of arcades, and controversies about violence all influenced the rise of a Gamer Nation, but the appeal of gaming chiefly came from the digital realm that gamers inhabited when they played: the imaginative spaces and stories that they were drawn to. Even as they physically sat at home in front of the television, players spent their recreational hours and had their formative experiences within digital game worlds. Such virtual, ephemeral spaces remain the least explored and least understood of cognitive territories. Gamer Nation is about the versions of America produced by programmer imaginations, computer chips, and joystick twitches. It is about "America" as simulation and digital story—and as a gamic code.

Deciphering that code represents the object of this study. *Gamer Nation* explores the depiction of America in video games from the 1960s to the present day. American-themed titles have proved to be some of the most successful video games across four decades of play. *Madden NFL*, *NBA Jam*, *Call of Duty*, *Grand Theft Auto*, and *Red Dead Redemption*—all framed around some key aspect of the American experience—have become hugely lucrative franchises. Alongside fantasy games depicting alien invasions, programmers have consistently reimagined the American experience as a game. Be it a national sport, a bustling cityscape, or an epic frontier quest, many titles have tackled aspects of American life. This project traces the *evolution* of a gamic America. It explores the "digital states" of America offered in gamer culture. It is here, the digital space that gamers reside in, where we have the potential to discover new things.

The term "evolution" suggests a sense of progression in digital projections of the United States, and certainly in terms of visuals, this is noticeably the case. In the 1960s and 1970s, titles that simulated American life, culture, and politics did so in rudimentary form. The first computer game, *Spacewar!* (1962), inspired by the Cold War, put just two tiny spaceships on screen. Early titles relied on symbolic representation, employing limited pixels and stick figures to denote periods of history and famous characters. With the advent of three-dimensional (3D) gaming in the 1990s, game worlds assumed new architectural sophistication. Titles replicated whole cities, such as New York and Los Angeles, albeit in a blocky style with hazy fog that disguised a lack of detail. Technological advances continually fueled a desire for greater realism and more elaborate simulation. In the 2000s, a fuller, more complete version of gamic America emerged. Software teams accessed historical archives and hired consultants to program accurate renditions of inner-city architecture. They modeled 3D sports cars and the faces of famous basketball players.

Across four decades of digital play, players found themselves visiting versions of America past, present, and future, while the wealth of titles produced a range of digital versions of America to explore and inhabit. In *Assassin's Creed III* (2012), set during the American Revolution, players assumed the role of Connor, a hunter with Mohawk heritage, navigating both the wilderness frontier and the colonial cities of Boston and New York. In *Grand Theft Auto V* (2013), players sat on Santa Monica beach, gazing out onto the Pacific Ocean, watching the waves while sipping a cocktail. Surfers strolled past, car engines purred in the background, and the local tennis coach stood across the street waiting to play a game. In *Fallout 4* (2015), set in 2287, gamers explored an irradiated New England wasteland through the eyes of Sole Survivor, emerging from fallout shelter Vault 111 alone and eager to find missing child Shaun, assisted by a personal computing device nicknamed Pip-Boy. Successive software teams forged a collective vision of American life founded on a range of archetypes, folklores, visual data, and stereotypes. This gamic version of America is of historical and cultural significance, and it is ripe for study.[18]

The study of these game worlds may, in some small way, help us make sense of a growing phenomenon in US culture: the power of simulation and fakery. The rise of replicas and facsimiles arguably marks post-1945 American society. Games intersect with film, television, advertising, and popular culture in the late twentieth century and early twenty-first century in promoting a visual fantasy of America: a world of superheroes, catwalk models, and fast cars. Mobile technology, the Internet, robotics, theme parks, the Disney town of Celebration, and the Las Vegas strip have all promoted a simulated, often idealized, American experience. Intriguingly, the rise of the video game in the 1970s coincided with a time of mass disillusionment with real American life, brought on by the Vietnam War, the Watergate scandal, and the energy crisis. People welcomed the simulated, hyper-real nature of video games for the same reason that they had once turned to dime novels in the 1870s or to cinema during the Great Depression: to escape reality and enter a simulated American experience. As one Sears representative explained about the appeal of *Pong*, "It is part of the consumer desire to get away from harsh realities."[19] Predicting the replacement of print culture with digital culture and the associated rise of a new "global village" of communication, media theorist Marshall McLuhan contended in the 1960s that "the games people play reveal a great deal about them," and I share his conviction.[20] Moreover, for some time now, scholars have noticed Americans' increasing fascination with simulation and the lure of fantasy. In 1988 French philosopher Jean Baudrillard spoke of America as a land of signs and symbols, a superficial realm marked by newness, facile pursuits, and disconnection. The rise of American-set video games might be taken as a natural extension of the "symbolic America" detailed by Baudrillard: an America consisting of images and icons, simply transferred to new media.[21]

Games also bring into focus the lived American experience. They explore what it means to be American from a new media and a new critical perspective. French commentator J. Hector St. John de Crèvecoeur asked in the 1770s on his visit to the East Coast, "What, then, is the American, this new man?" Video game designers have explored a similar conundrum two centuries on and, in the process, offered their own definitions of a new kind of digital citizen. Since their inception, video games have critiqued facets of American society, culture, and politics. In their simulations of Cold War attack, financial empire-building, and Wild West high noons, they have imparted stories of success and failure. Video games have reflected the anxieties and the opportunities of everyday life, and they have spoken (or, rather, played) to the American zeitgeist.

As gaming has become more popular, the cultural relevance and impact of video games has also increased. In 2017–18, several blockbuster titles directly tackled the American condition. Ubisoft's *Far Cry 5*, a first-person adventure game about cults, community, and conflict in Montana, highlighted the ability of games to explore deeper notions of place and belonging. The title explores rural impoverishment in the West, showing how disenchanted Americans could easily be drawn to the promises of salvation offered by maverick leaders and how the community could be divided by those loyal to different versions of the American Dream. Quantic Dream's *Detroit: Become Human* depicts the story of three sentient androids residing in futuristic Detroit. The game takes on issues of slavery, sentience, and individual rights through the lens of cyborg citizens. It makes frequent allusions to African American struggles in Detroit in the 1960s, with one android called Marcus even delivering his own "I Have a Dream" speech, a nod to Martin Luther King Jr.'s call to action, first delivered before amassed crowds in Detroit in June 1963, then later, more famously, in Washington DC. Set for release in October 2018, Rockstar's *Red Dead Redemption II*, an adventure about the outlaw Arthur Morgan and set on the American frontier in 1899, promises the most authentic exploration yet of the Western diorama.

Of course, games do not always require classic American settings to speak to the American condition. Epic Games' *Fortnite*, a multiplayer shooter game released in 2017, alludes to a range of American themes and ideas despite its fictional island setting and stylish cartoon graphics. A free-to-play version of the title emerged as a phenomenal success in 2018, attracting over 100 million players worldwide. In creating the game, Epic, a North Carolina company founded by developer Tim Sweeney, exploited a range of cultural, business, and technological trends floating in the ether. *Fortnite* offers an American dystopia that speaks to the 2010s, presenting a (game) world torn apart by climate change and meteoric storms, fractured and individualistic in nature. Epic then filled that dystopia with a popular foe drawn from late twentieth- and early twenty-first-century fiction: the zombie. In the late 1960s, cult director George A. Romero shot the low-budget horror film *Night of the Living Dead* (1968), a black-and-white zombie film that served as subversive commentary on the tumult of the period, in particular the violence of Vietnam and clashes in race relations. Romero followed it with *Dawn of the Dead* (1978), in which survivors defended a shopping mall from mindless materialist zombies. Zombies first appeared in video games in the early 1980s, as fast-moving pixel enemies in US Games' maze game *Entombed* (1982). In the 1990s, thanks to titles such as Capcom's *Resident Evil* (1996) and Sega's *House of the Dead* (1997), zombies came to dominate the survival-horror subgenre. By the 2010s, zombies were one of the quintessential video game enemies; several gaming franchises (including *Call of Duty* and *Red Dead Redemption*) incorporated dedicated "zombie levels."

Fortnite also relied on another modish convention: the battle royal. The concept of the battle royal—an epic fight for survival—can be historically traced to the opening scene of Ralph Ellison's *The Invisible Man* (1952), in which a blindfolded black child fights for the entertainment of white elites. In the mid-1990s, the battle royal appeared on television in the form of sensational wrestling events. The World Championship Wrestling (WCW) hosted an annual series of bouts called *World War 3*, involving up to sixty fighters. In his novel *Batoru Rowaiaru* (1999), Japanese author Koushun Takami refined the concept, adding wider dystopian themes, with junior high school students forced by an authoritarian government to fight to the death (a film version directed by Kinji Fukasaku was released to critical acclaim). Suzanne Collins's *The Hunger Games* trilogy (2008–10) set the fight in the dystopian realm of Panem, located in future North America, with district lottery winners (labeled Tributes) coerced into spectacular televised, hand-to-hand conflict by the repressive Capitol. This American version of the battle royal concept—with its notions of survivalism, individualism, competition, bravado, revolution and even romance—was attractive to audiences. Epic's *Fortnite* allowed gamers to not just watch, but play their own hunger games (minus the aspects of real-life sacrifice). With its Pixar-like cartoon style, Epic's title exaggerated aspects of play and performance, while downplaying the mature content.

Furthermore, Epic tapped the increasing popularity of microtransactions, whereby players can purchase specific game enhancements such as "skins" (avatar outfits). Epic's title proved so popular that worries over the harmful effects of gaming resurfaced, specifically issues of addiction, with critics voicing concerns similar to those of Ronnie Lamb in the early 1980s. *Fortnite* highlights the degree to which successful video games not only offer immersive play experiences, but articulate popular American ideas.

This book thus looks at how games have simulated and begun to reframe the American experience. Rather than dwell on binary questions (with more-than-binary answers), such as "Do games lead to violent behavior?" it instead considers the subtleties of game content. Rather than narrate the history of the video game industry in the United States, it probes the imagined worlds and stories produced by it. While issues of reality versus fiction regularly occur in discussions of gaming, I am not concerned here with historical accuracy in gaming (an already old debate), but rather with what the fusion of real and fiction co-creates. In Game Studies terms, I am more interested in representation in (and the narrative geography of) games, than in their rules or internal dynamics. This book is a cultural history of game worlds. It looks at the "American world" through the game itself, exploring the new terrain of an imagined and digitally coded nation.[22]

Understanding Video Games

Despite their near half-century of use, video games remain only partially understood. Popular histories of the American video game industry first appeared in the 1980s thanks to hobbyist enthusiasm, in the guise of trade magazine summaries and Leonard Herman's self-published *Phoenix: The Rise and Fall of Home Video Games* (1984). Journalistic accounts followed in the 1990s, with J. C. Herz's *Joystick Nation* (1997) and Stephen Poole's *Trigger Happy* (2000). The academic discipline of Game Studies emerged in the late 1990s, with Espen Aarseth's *Cybertext: Perspectives on Ergodic Literature* (1997), which tackled new forms of communication found in MUDS (multiuser dungeons); Justine Cassell and Henry Jenkins's *From Barbie to Mortal Kombat: Gender and Computer Games* (1998), which addressed gender issues; and Janet Murray's *Hamlet on the Holodeck: The Future of Narrative in Cyberspace* (1997), which explored the notion of games as interactive theater. Scholars initially focused on how to interpret games, and specifically their similarity to and difference from other media. The issue of ludology versus narratology shaped early debate, with Gonzalo Frasca and

Jesper Juul arguing the need for a play-centered interpretation. *Man, Play, and Games* (1958) by Roger Caillois manifested as a popular reference, thanks to the French philosopher's decoding of "play" around themes of competition, chance, imitation, and vertigo (or thrill), all aspects of relevance to the video game. Meanwhile, Mary Fuller and Henry Jenkins compared Nintendo's *Mario* games to New World travel writing. Game Studies helped reveal the distinctiveness of games: their procedures, patterns, and potential.[23]

As the field of Game Studies advanced, scholars explored the broader cultural and historical aspects of gaming. The function of video games as cultural products garnered interest. For Garry Crawford and Jason Rutter, games existed as "cultural artefacts which are given meaning and position through their production and use," while Simon Egenfeldt-Nielsen et al. commented on how "the idea of games as reflections of cultural themes remains an interesting but under-explored idea," an idea that shapes much of this book. Scholars began to tackle how video games depicted the past. The online discussion forum Play the Past, along with Zach Whalen and Laurie Taylor's *Playing the Past: History and Nostalgia in Video Games* (2008) and Matthew Wilhelm Kapell and Andrew Elliott's *Playing with the Past: Digital Games and the Simulation of History* (2013), outlined the concept of digital history. In *Digital Games as History: How Videogames Represent the Past and Offer Access to Historical Practice* (2016), Adam Chapman provided valuable theoretical viewpoints on historical gaming and explored the potential of games to offer "history beyond the academic word." In the 2010s, scholars began to map out America's distinctive relationship with gaming in terms of cultural and social history. Through the lens of technological progression, Caroll Pursell depicted a range of recreational stages leading to the advent of video games in *From Playgrounds to PlayStation: The Interaction of Technology and Play* (2016). Michael Newman documented 1970s gaming in *Atari Age: The Emergence of Video Games in America* (2017), highlighting many of the cultural contradictions and confluences that marked the first decade of mass digital play. Carly Kocurek detailed arcade life in the early 1980s and the rise of the "technomasculine" gamer in *Coin-Operated Americans: Rebooting Boyhood at the Video Game Arcade* (2015).[24]

Gamer Nation is similarly a work of cultural history, but my aim is somewhat different from these titles, in that I explore America's relationship with video games through specific software titles. I look at how the games themselves depicted American culture, values, and history. In a sense, by highlighting the visions of America offered in a range of titles across several decades, I offer a more game-centered analysis.

While the overall trajectory of Game Studies has been to assert the distinctiveness of video games and their study, games might still be treated as part of the wider process of literature, film, and performance in framing popular culture. Rather than tackle games in isolation, I situate video games alongside other media in molding and framing the American condition. Using a persuasive myth-symbol approach, Henry Nash Smith in the 1950s highlighted the cultural power of nineteenth-century and early-twentieth-century popular Western literature, particularly the dime novel. In *Virgin Land: The American West as Symbol and Myth* (1950), Smith documented the prevalence of powerful symbols and myths in US society, with certain imagery, such as the violent frontier, endemic in mass culture.[25] Like dime novels a century prior, video games fit within the much-denigrated bracket of "low culture" that Smith was most interested in—a category often neglected by the scholarly lens. In fact, early video games resemble dime novels in their repetitive, simplistic, action-based storytelling; their targeting of mainstream audiences; their reputation for dubious content, such as killing, violence, and titillation; and the claims made of their damaging influence on society (in other words, their ability to incite moral panic). Like dime novels, video games are part of a longer process of cultural myth-making and popular storytelling about America. Despite their distinctive iconography, video games draw heavily on historic symbols.

Approaching a Digital America

Three initial ideas shape my approach to understanding gamic versions of America: the power of immersion, the "magic circle" blur, and the motion of "reframing." Although scholars have yet to fully understand the mechanisms involved, all agree that a crucial difference exists between playing something and watching it. Video games differ from other media (such as cinema) by the ways in which they involve the audience. By inviting the player to "act," they possess an immersive power arguably far more potent than the voyeurism of film. What happens when stories, messages, and instructions are inevitably more involving? Films in the 1980s such as *Rambo* (1982) effectively combated the phenomenon of "Vietnam Syndrome" in American society by tackling issues of shame, redemption, and remasculinization, and even provided a substitute victory. What did video games offer American society in the late twentieth and early twenty-first centuries? What narratives did players experience, and crucially what narratives did they co-create? What versions of America were players immersed in, and what landscapes were made?[26]

Games also question notions of reality, identity, and space. Fredric Jameson wrote, "American literature never seems to get beyond the definition of its starting point: any picture of America is bound to be wrapped up in a question and a presupposition about the nature of American reality."[27] Exploring the detective fiction of Raymond Chandler, Jameson touched on a much broader point regarding immersion: the puzzling nature of reality and the desire to "know" what American life is about. In 1938, Dutch scholar Johan Huizinga coined the term "magic circle" to describe the distinctive and highly fictive arena of game worlds: the enclosed reality (or "life") of games.[28] Huizinga referred to the boundaries and rules of tennis courts and card tables, not the game worlds of Atari *Pong* or iPhone *Solitaire*. In the 2000s, however, Games Studies scholars explored the concept of the 'magic circle' in relation to computer games to describe self-enclosed game worlds marked by unique rules and concepts. Edward Castronova posited that the 'circle' might be broken, potentially even dissolved.[29] The two strands of "what is the nature of American reality" and "what is gaming reality" combine in this work. Usefully applied to American-themed games, the magic circle is a coded, controlled, and fictionalized version of a specific US place. Players enter the game-world, the simulated reality, knowing its borders and differences from the outside, or real, world (in fact, like theme parks, games often preserve a knowing difference or uncanny aspect). However, what happens when the two worlds influence each other, bleed into one another, and co-create one another? Are video games reprogramming the American experience as a new and heavily simulated reality?[30]

Nobody can precisely surmise how video games (or films or literature, for that matter) reshape how we think or frame our forms of politics, society, or culture. At the same time, video game culture clearly became part of American life in an increasingly significant way from the 1970s onward. Immersion in a magic circle of an imagined America likely led to some instances of seeing the "real" America in new and different ways. The fake or simulation inevitably shapes our view of reality, as Umberto Eco found on a trip in the 1980s to Disney's version of the American South prior to embarking for the real New Orleans. Perhaps, the framing happens when an international gamer plays Rockstar's *Grand Theft Auto*, establishing his or her "original" America before disembarking in New York as a fresh immigrant; the initial wanderings in the *GTA* game world replace the first vision of the Statue of Liberty as the gamer's introduction to America. Perhaps the forty-hour-a-week player finds himself viewing his surroundings differently, confusing game time with out-of-game time. Perhaps after completing the apocalyptic journey of *The Last of Us* (2013), about a nation slipping into decay, the player thinks differently about the fate of the real America outside. Video games not only signify a key aspect of contemporary American culture, but in their depictions of cities plagued by crime and troops gunning down terrorists, actively play with the way people relate to America itself. Digitally constructed realms toy with our mass historical and cultural understanding. As Timothy Welsh contends, there are clearly (and often quite complicated) "flows between on-screen and off-screen." Games reframe popular depictions of America past, present, and future, and they may refocus our collective lens.[31]

Chapter Plan

This book provides an analysis of video games within American culture; their presentation of America past, present, and future; and their potential to reframe American experience. I focus on the depiction of the United States in a range of selected games, exploring problems, possibilities, and impacts associated with the media.

In the first chapter, I look at the emergence of a gamic America. I link the rise of computer and video games in the 1960s and 1970s with a search for new technological frontiers, the lingering influence of the amusement park and the pinball arcade, and the genesis of the computer age. I look at how early games, through stick figures and simple baseball games, depicted everyday American life. The concept of the "frontier" is then explored in two well-regarded computer games, the education title *The Oregon Trail* (1971+), which taught schoolkids about westward expansion, and *Civilization* (1991), a Sid Meier strategy title about conquest. Both titles cast the player as an avid explorer and discoverer. Both titles provided an interactive and immersive national past to negotiate and an early version of a gamic America to explore.

The notion of games exploring the American past continues in chapter two, where attention shifts to games that depict one specific period of history: the nineteenth-century American West. Tackling mostly arcade titles from the 1970s and 1980s, the chapter highlights the synthesis of classic folklore about the "Wild West" with action-based shooting games. Arcade machines such as Midway's *Boot Hill* (1977) celebrated the cowboy persona and glorified the romance of the gunfight and the showdown.

Video games are not restricted to resurrecting the distant past; they have also depicted America in the post-1945 era. In chapter three, I look at games published in the 1980s that coincided with the second Cold War and the political rise of Ronald Reagan. Titles such as Atari's *Missile Command* (1980) and Coleco's *WarGames* (1984) explored the potential ramifications of nuclear conflict in the period. By highlighting the unwinnable nature of nuclear war, games offered poignant political commentary and even protest.

The theme of coping with impending disaster continues in chapters four and five, where I tackle the relationship of video games to the 9/11 terrorist attack. I explore "premonition titles" that foresaw the act of mass violence and look at the long shadow that 9/11 cast over American culture, including its effect on video games. I consider the renegotiation of national trauma and the rise of the "patriotic soldier" in the "War on Terror" video game genre. Do games of the 2000s play a similar role as *Rambo* films of the 1980s in assuaging national trauma?

While New York City provides the real-world location for chapters four and five, attention shifts to the West Coast for chapter six, specifically to Los Angeles. This chapter considers Rockstar's recreation of the West Coast metropolis in its acclaimed (and controversial) *Grand Theft Auto* series. Published in 2013, *GTA V*, the most recent iteration of the franchise at the time of writing, provides gamers with an incredibly immersive, photo-realistic, and road-accurate version of the City of Angels, renamed Los Santos, in which gamers can "play criminal." It depicts America as a deranged, disturbed, and yet somehow alluring dystopia.

For every imagined dystopia, there exists a utopia. In chapter seven, I tackle a title that challenges traditional notions of gaming, but equally offers one of the most immersive realms of play in the 2000s. Released in 2003, *Second Life* by Linden Lab provides an online, virtual landscape for Americans (and others) to forge a second, virtual life. While many members of *Second Life* choose fantastical abodes to inhabit, a number of "New World colonists" see themselves as founding fathers of an America 2.0 and create facsimiles of US architecture, values, and living. While now graphically dated, *Second Life* points toward the potential future for gaming—what forms a gamic America might take ahead, with a new frontier of virtual reality looming.

Together, the seven chapters narrate the evolution of video game depictions of America, from stick-like symbolic presentations in the 1970s through to sophisticated, photo-realistic depictions in the 2010s. The book highlights how games have played with the past and perpetuated cultural myths of freedom and the frontier, how they have reflected late-twentieth-century fascinations with crime and the American city, and how gaming culture has forwarded conflict and killing as "what you do" in fictive America. It also highlights the value of video games in the exploration of both optimism and pessimism surrounding the American future. Alongside familiar takes on history, video games have the potential to offer new understandings about American life.

Games and New Frontiers

In 1964, designers Charles and Ray Eames produced the striking documentary *IBM at the Fair*. The twenty-minute reel provided a visual document of life at the IBM Pavilion, a huge, white, egg-shaped construction at the 1964 New York World's Fair. Employing time-lapse photography, Eames presented a world of perpetual motion powered by new technology. As if gaining their kinetic energy from the electronics, businessmen raced at hyperspeed among corporate exhibits, pressing buttons on giant computers. Children played on a bewildering electronic, gravity-based ball game. For the first time, people seemed connected to electronic machines in a fundamentally playful way. A giant multiscreen auditorium delivered the Eames production *Think*, a lavish artistic peek into the future marked by an array of flashing images delivered at high speed. With frenetic supremacy, *Think* bombarded the viewer into submission to a future of multiscreen data and a life determined by technology. Visitors peered at the IBM stand through scopes, cameras, and film lenses, as if staring directly into the future. That future appeared high speed, social, visual, and like a computer game.[1]

The same kind of ideas could be found across the rest of the World's Fair. New York looked to a new horizon of technological mastery and prowess: a world of space cars, automation, computers, and play. Dedicated to "Man's Achievement on a Shrinking Globe in an Expanding

Universe," the twelve-story Unisphere, constructed from stainless steel, dominated the skyline, a visual marker to a smaller world constructed and bound together by modern technology. Designed by Walt Elias Disney, General Electric's Carousel of Progress depicted an American home transitioning from the nineteenth century to the 1960s and beyond, with new gadgets making family life easier and more fun at every stage. The saccharine, upbeat tune "There's a Great Big Beautiful Tomorrow," penned by the Sherman Brothers, played in the background. As a "future playground," the World's Fair constructed an America marked by bountiful electronic energy.[2]

Such technological optimism coincided with the concept of a revitalized American frontier. At the Democratic National Convention in Los Angeles on 15 July 1960, presidential nominee John Fitzgerald Kennedy sketched out his sense of a country on the brink of something new: "We stand today on the edge of a New Frontier—the frontier of the 1960s, the frontier of unknown opportunities and perils, the frontier of unfilled hopes and unfilled threats . . . Beyond that frontier are uncharted areas of science and space, unsolved problems of peace and war, unconquered problems of ignorance and prejudice, unanswered questions of poverty and surplus."[3] Kennedy fused classic frontier rhetoric with his own desires for domestic and social reform. He sought to capture the emotional power of American frontierism and tap into a nation-building mythology for his cause. Kennedy updated the frontier concept to include overcoming the social challenges of the day. The new frontier welcomed breakthroughs in science and technology in the guise of space travel. Caught in a Cold War climate, with escalating tensions between the Soviet Union and the United States, America embraced the frontier again, while on the precipice of nuclear war.

Translated into the triumphalist lexicon of the World's Fair, Kennedy's frontier operated as a largely cohesive fantasy made up of progressive white cities, space colonies, and patriotic endeavor. Since the Great Exhibition in Hyde Park, London, in 1851, World's Fairs and Expositions had showcased the latest industries and inventions, and entertained lofty future prospects. The New York World's Fair, along with its precursor, the Century 21 Exposition, held in Seattle in 1962, served as conspicuous propaganda pieces for a revitalized country. Both events highlighted what could be achieved with a technologically advanced liberal–capitalist nation. Both celebrated the ideological construct of a "new America." Rife with technological optimism, the Seattle Exposition featured a range of monuments to the new frontier, including the Space Needle observation tower and a Ford-designed space car. A huge 100-person-capacity "bubbleator" (a futuristic-looking glass elevator) transported visitors to a "future America," where fallout shelters gave way to glossy images of Marilyn Monroe. At the World's Fair in New York, exhibits such as the Unisphere, the Carousel of Progress, and *Think* sold the new frontier. Rockets at NASA's New York Space Park projected upward into the sky. True to the tradition of World's Fairs as "future playgrounds," the expositions in both Seattle and New York painted a technological future marked by optimism, imagination, and fun.[4]

For many participants at both fairs, one thing was abundantly clear: the new frontier of the 1960s was to be powered by computer technology. Computers made the small calculations that allowed the big machines to function. They evaluated data for space flights, predicted troop maneuvers in war games, and made home appliances work. The Seattle and New York fairs factored the computer into their space-age, atomic-powered white cities, and projected images of an automated leisure society. Similar to Disney Imagineers, World's Fair exhibitors forged fantasy lands of future prosperity and utopian living. General Motors' Futurama II provided visitors with a tour of three-dimensional models of colonies prospering on a variety of frontiers, from the deepest oceans and densest jungles to ice and lunar settlements. More than twenty-six million people took GM's "ride into tomorrow." Incorporated into such visions, early computers signified a "better life" marked by leisure, wealth, and abundance. At the Seattle fair, a UNIVAC computer powered "the library of the future," while at the New York fair, Bell Telecoms demonstrated an early computer modem, along with a computer-based international pen pal finder, nicknamed the Parker Pen.[5]

Shown across nine screens and employing a total of fourteen projectors, the Eames piece *Think* visually underlined the centrality of computer logic. Present at both fairs, IBM positioned its machines as at the processing core of the new technologically fueled frontier. Dating back to the 1910s, IBM was a well-proven pioneer in technologies designed to assist businesses. By the 1960s, the corporation had introduced one of the first mass-produced computers, the IBM 704, and developed

FORTRAN, an early programming language. It had also assisted the government with the space program, including supplying guidance computers for NASA's Saturn rockets.[6]

IBM demonstrated that computers were about more than business and information: they were about play. At the IBM Pavilion at Seattle's fair, a child's maze presented logic code as a traditional park-style game of navigation, while an early computer played tic-tac-toe with guests, and ShoeBox demonstrated voice-recognition technology. At New York, an electromechanical puppet show called "Puppets in the Park," reminiscent of the older amusement staple of Punch and Judy, relayed the story of computers through a Sherlock Holmes–style adventure. The media enthused over the "delightful" detective show.[7] Visitors to the Pavilion wrote dates on cards; computers, using an early handwriting recognition system, translated the cards and then responded with date-related facts. Thrilled crowds found computers answering their questions. Information appeared a game. Human and machine connected, for most participants for the first time, around a situation of play. The IBM Pavilion thus introduced visitors to the wonders of the computer age through the conduit of games.[8]

First Glimpses of a Gamic America

The idea of playful computers operating on a new frontier proved to be a romantic image of the World's Fairs of the 1960s. However, the connections of frontier, play, and computer technology being made in Seattle and New York were not isolated to the exhibition spaces; they reflected energies outside of the fairs. Staff familiarizing themselves with IBM mainframe computers and DEC PDP machines created early computer programs for fun. University students, hijacking faculty resources for their own recreational uses, developed basic computer games. Games also emerged as a product of the US military–industrial complex.[9]

In 1958, William Higinbotham created a basic tennis simulation on a Donner Model 30 at Brookhaven National Laboratory. A former Manhattan Project Los Alamos nuclear scientist, Higinbotham had worked on timing circuits for the first atomic bomb. Based at Brookhaven, he designed *Tennis for Two*, essentially a game about a game, and the first attempt at putting a recreational sport on screen. Using a DuMont oscilloscope as the display, the game showed a tennis ball on a court, with players using purpose-built aluminum controllers as racquets. Each controller featured a single red button for hitting the ball and a rotating knob to control the angle of shot. Designed for the annual Brookhaven exhibition, Higinbotham hoped that his game would attract crowds and invite broader interest in scientific work. Visitors loved it, to the extent that it reappeared the following year. *Tennis for Two* emerged from experimentation, playfulness, an interest in simulating American sport, and a desire to advertise American science.[10]

In the popular consciousness, early computer games came to embody the "soft side" of the new technological frontier. The product of free time and hobbyism, computer games highlighted the other side of serious application. They introduced people to new technology, and they entertained. As the notion of "play" suggested idle fun rather than serious endeavor, critics underestimated the impact of these early games, which limited understanding of their cultural meaning. Both public and media response to computer games was muted. With the first games designed and played by only a few people, it was no surprise that "the human race at large paid no notice to the fact that video games had been born."[11] While the nation gasped at the first manned moon landing in 1969, few heard of the release of the first arcade machine, *Computer Space* (1971). Right from the start, games were sidelined as a frivolous fragment of the new digital realm that the United States entered in the 1960s. Gaming proved to be an underestimated driver of the new digital age.

This fringe aspect disguised the number of games that had serious origins, applications, and repercussions. Rather than being a sideline to the serious business of industry and innovation, the ludic often chimed with visions of the American future and helped determine the form and direction of the US computer industry. Play was intimately connected with the most serious of projects at scientific laboratories and military sites. Programmers employed games to explore all kinds of potential scenarios, from space fuel projections to military warfare. As Caroll Pursell notes, "The birth of video games in the nation's university laboratories during the Cold War meant that they were never far from the massive commitment to military research that dominated those years."[12] In late 1952, George Gamow at Johns Hopkins University Operations Research Office developed a chess-based simulation of military combat. A computerized version followed just four years later, a project headed by Richard Zimmerman. Working on the UNIVAC 1103A from 1956 onward, Zimmerman's Carmonette (Combined ARms Computer MOdel) served as a template of combat for the US military in the 1960s and early 1970s.[13] The RAND Corporation developed nuclear war games for the US Air Force during the same period. Games also factored into the evolution of the computer industry and shaped ideas about programming, peripherals, and interfacing. The first example of artificial intelligence heralded from the IBM Poughkeepsie laboratory, when in 1956 an IBM 704 learned from its mistakes while playing a game of checkers. Programmers used games to explore the potential of new designs of computers. Pushing the boundaries of graphical ingenuity and highlighting the possibility of artificial intelligence, games showcased the power of computer hardware.

Games also played a role in the emergence of a new technological society. Arcade games became early symbols of a fledgling computer society and a new Gamer Nation. In a relatively short period of time, the first purpose-built game machines appeared in bars, arcades, and homes. Atari's release of the table tennis game *Pong* (1972) marked a new gaming era. Texas Instruments, Atari, and Apple introduced the first affordable home computers in the late 1970s. Advertising their diverse applications—from teaching typing and French to mortgage and stock analysis—microcomputer companies encouraged people "to explore the varied uses and pleasures of this new technology." The American public tried coding for the first time, learned computer languages such as BASIC, and created their own game titles. Michael Newman notes how computers quickly moved from novelty status to familiar household objects, with "games and play always at the center of this."[14] The gaming revolution necessarily involved real spaces. Laboratory cupboards, computer rooms, bedrooms, garages, arcades, and malls served as the physical spaces of the electronic frontier. Early programming gave rise to a new youth and college culture centered around computers and games: a gamer generation. Closed circles of programmers and gamers—labelled nerds and geeks by those outside the new gaming culture—banded together. The new American frontier of digital play happened in the basements of MIT and the bedrooms of college. The digital frontier gradually forged and reassembled real, as well as virtual, social spaces.

While pioneering new forms of entertainment, the gaming frontier of the 1960s and 1970s inevitably drew on a range of influences. A Gamer

Nation emerged from a variety of impulses, some unique to the technology itself, others from established recreational pastimes and traditions of play. Looking at the interfaces of play with twentieth-century work, technology, and family, Caroll Pursell notes how "play has been deeply embedded in the American experience."[15] Despite the popular image of gamers in the period as boys in dark arcades or bedrooms, gaming was never completely isolated. Video games brought people together and reflected the rise of technology as a hobby.

Early computer games drew on notions of narrative heavily embedded in popular culture. Programmers harnessed familiar literature tropes and storytelling techniques. Focused primarily on action, arcade games provided the bare bones of traditional storytelling; they assumed players already instinctively knew the story. Games harnessed a familiar literature based around the tropes of fantasy and discovery. Exhibiting clear travel-writing influences, early games told spatial stories based around the activities of mapping and exploring. Jumping across quicksand and dodging scorpions, players imagined themselves as intrepid colonial explorers in Activision's *Pitfall!* (1982), a side-scrolling adventure game inspired by the Steven Spielberg movie *Raiders of the Lost Ark* (1981) and classic Tarzan stories.[16]

Games also drew extensively on the medium of film. In the 1880s, Eadweard Muybridge recorded a galloping horse in Palo Alto, California. His series of photographs revealed what racehorse owner Leland Stanford had long suspected: the "unsupported transit" of equines at speed, with all four hooves lifting off the ground.[17] In the early years of celluloid, film was employed to capture all kinds of movement, from tennis to trains, from chasing dogs to dancing skeletons. The motion picture was fundamentally about motion. Another highly visual medium, early video games similarly focused on movement. Replicating the source material and lens of early film, Atari's *Pong* depicted a game of table tennis and Midway's *Wheels* (1975) showed a car race. Out of the twenty-eight games released on the Magnavox Odyssey console between 1972 and 1973, twenty were based around motion. Dots flew across the television screen, awkwardly mirroring the trajectories of balls, bullets, and missiles.

Like early film, early arcade games represented incredibly short cycles of entertainment. Technical challenges impeded the storytelling abilities of both formats. As with initial generations of film, games lacked realistic sound and instead used a variety of beeps to suggest a greater aural experience. Furthermore, neither format offered sophisticated theater from the outset. However, enforced limits also led to creativity. In the case of film, in *A Trip to the Moon* (1902), one of the first science-fiction films, George Méliès used mechanical scenery and substitution splicing to great effect, while Higinbotham's *Tennis for Two* impressively mimicked gravity.

Early filmmakers longed for interactivity with the audience and to cater to new fans. The first action movie, Edwin S. Porter's *The Great Train Robbery* (1903), borrowed a range of Wild West conventions to simulate an attack on a locomotive. The film was notable for its location shooting, new techniques, and quick pace. In the final scene, Justus D. Barnes points his revolver at the cinema audience and fires. Audiences recoiled in shock. While film settled into a largely noninteractive experience, directors still experimented with methods to involve their audience to a heightened degree. Three-dimensional movies using specially adapted auditoriums proved popular in the 1950s, a classic example being the ant movie *Them!* (1954). In the 1960s, horror director William Castle used electrical buzzers beneath auditorium chairs to elicit shocks from the audience as they watched his movie *The Tingler* (1959).[18]

Game designers, by contrast, sought to make an interactive medium more cinematic. Reflecting the dominance of film in popular culture, game design studios consistently sought film-like qualities, especially in terms of visuals. Programmers longed for cinematic production values. In the 1970s, 4-kilobyte computer memories produced stick figures, at best. In the 1980s, as graphics chips progressed, a number of titles began to emulate film. Atari versions of popular movies such as *The Empire Strikes Back* (1980) and *Ghostbusters* (1984) condensed film narratives into a number of action set-pieces. Cinemaware, founded in Burlingame, California, published a range of home titles based on film genres, including World War– and Cold War–inspired games. Dubbed "interactive movies," laserdisc games such as *Dragon's Lair* (1983) by Cinematronics played as a series of cut-scenes and mimicked film closely. At the arcade, audiences played *Mad Dog McCree* (1990) by American Laser Games, an escapade not that far removed from *The Great Train Robbery*, but when faced with the villain, the players got to shoot back.[19]

Critical responses to early games mirrored responses to the advent of film. Nickelodeon theaters in the 1900s attracted a range of customers, but as a place for the working class and youth to congregate, such venues incited moral panic. A similar response occurred with video games. As Carly Kocurek details, 1970s arcades became linked with fears about gambling, criminality, and poor behavior, and violent titles such as *Death Race* "became a lightning rod for a broader uneasiness about games."[20] The press condemned both new media as morally corrupt. The poorly lit physical spaces of both the cinema and the arcade appeared to nurture deviant behavior. Fear of the damaging effects of video games also reflected broader concerns about youth culture and rebellion. Campaigns such as the one led by Long Island PTA member Ronnie Lamb asserted the danger of new media to a nation's youth, and according to Stephen Kent, "helped to sour the public's perception of video games."[21]

Often presented on small cathode-ray tubes, video games also had strong ties with television. In 1972, Magnavox released the first home console, the Odyssey, designed by a team led by Ralph H. Baer, an electronics engineer. Baer had first conceived of a television that played games in 1951, but he didn't begin working on a prototype until 1966, in conjunction with Sanders Associates. Baer sought to connect games literally as well as figuratively with the home television set.[22] The notion of game channels replicated television channels, with manufacturers in the 1970s widely promoting their products as "telegames." Newman notes how video gaming provided "a new use for a familiar technology," stating that "television was at the core of the new technology's identity."[23] Games reinforced the primacy of home entertainment built around the television by using it as a display. However, for Atari founder Nolan Bushnell, the television was never more than "a cheap display mechanism" for his video games, a simple means to an end: "I didn't see television as relevant in what I was trying to accomplish other than as vehicle."[24] Games also began a journey through the same arc traveled by film: from purpose-built viewing, in cinemas or arcades, to watching at home. The welcoming of home consoles reflected the broader infiltration of technology into the home. Gaming consoles symbolized the modern, leisure-focused home environment.

Computer games also drew on their ludic heritage. Games reproduced games. For Bushnell, "games have been culturally part of the human situation since cavemen. As new technology emerged over time, people have always been receptive to new forms of gaming."[25] Programmers fashioned digital versions of popular board games such as chess and national sports such as baseball and football. The Odyssey came with real dice, poker chips, and fake money.[26] Plastic overlays transformed the machine into a miniature casino. In addition, video games drew on popular trends in postwar gaming, turning strategy titles such as *Risk* and role-playing games such as *Dungeons & Dragons* into computer titles. Giving old games new life, computers expanded the popularity of gaming. Recognizing consoles as a lucrative extension of the entertainment market, American toy companies funded the technology, with Milton Bradley investing in a fledgling game console, the Vectrex, in 1983.

The transformation of mental puzzles into a binary world proved relatively simple. Logic-based titles translated well into computer languages based around ones and zeros. Data-based games suited computers designed to process information like advanced calculators. Ludic heritage valued abstraction and symbolism. Games such as checkers or chess hardly resembled real life and thus never demanded graphical complexity or aesthetic realism. Other games, such as tennis, could be broken down into basic physics. Early video games carried a ludic focus, as seen in the most successful titles of the 1970s, *Pong* and *Breakout* (1976). Players adjusted well to these simple digital recreations of popular games.[27]

Furthermore, game development drew on the traditional US amusement industry. Bushnell worked as an operative at Lagoon Amusement Park in Utah before establishing Atari. He later reflected, "Lagoon was pivotal for me and gave me an understanding of the economics of the arcade business."[28] Since the Midway Plaisance at the World's Columbian Exposition in Chicago in 1893, World Fairs and Expositions had given room to dedicated amusement areas. Across the country, dime museums and penny arcades in major cities catered to a mostly working-class clientele and featured early coin-operated devices. George Tilyou took the concept of the Midway Plaisance and created Steeplechase Park at Coney Island, New York, in the 1890s. Dubbed the "nation's playground," Coney Island offered a world of circus-like play. Coney games included prize-fighting, shooting, and gambling. In Anaheim, California, Disneyland provided more family-friendly amusements, along with themed worlds, such as Frontierland, a Disneyfied version of the Wild West. Walt Disney promoted the idea of innocent play and child-like amusement. Along with Coney Island and Disneyland, Las Vegas represented another key American landscape of play. In the 1950s, a range of themed casinos emerged, including the New Frontier, all featuring gambling tables, slot machines, and showgirls. Video games added to the panoply of technologies, strategies, and theming employed in mass entertainment. As Newman explains, video games provided "the continuation of established practices and ideas about spaces of amusement."[29]

Companies that designed and assembled attractions for amusement parks and casinos became early converts to video game manufacture. Most American pinball companies became gaming companies. Used to manufacturing electromechanical pinball machines, the jump to designing arcade cabinets proved relatively small. As Pursell explains, "The electromechanical controls of the 1950s and '60s were replaced with circuit boards and digital displays of the '70s."[30] Founded in Chicago in 1958 by Henry Ross and Marceline Wolverton, amusement manufacturer Midway shifted to game development in 1973, while Williams, a prominent pinball company, also founded in Chicago, entered the arcade machine market with *Paddle Ball* (1973), a game derivative of *Pong*. Based in Japan, Sega Enterprises manufactured the electromechanical shooter game *Periscope* (1965) before properly entering the video game market in the early 1980s. Pinball machines and video games came to occupy the same geographical spaces, standing alongside one another at arcades, shopping malls, and amusement parks.

The influence of the amusement industry could be recognized in game content. Programmers created game worlds that sought to replicate the colorful architecture and adrenaline rides found at Coney Island and Disneyland. Popular titles such as *Duck Hunt* (1984) and *Carnival* (1980) both replicated classic fair amusements. Games provided a sense of visiting a miniature amusement park in the home.[31]

Video games thus emerged in the 1960s and 1970s as part of a historic and expanding entertainment frontier that coincided with increased leisure time in postwar America. Welcomed into the home, game consoles joined the television, board game, and radio as recreational entertainment. They connected with amusement landscapes such as Coney Island, Disneyland, and Las Vegas: places that resounded with frontier ideals and that for years had grouped together technology, play, and recreation. Video games merely extended the parameters of play.

Traditional media, including film, books, and television, all reflect and shape American identity and culture. From Harriet Beecher Stowe's

Uncle Tom's Cabin (1852) to the TV sitcom *Friends* (1994–2004), fiction and story have informed American life. How did early computer games present America, and how did they frame American culture? Limited by processing power, memory, and pixel quotient, early games struggled to offer sophisticated storytelling. In stark contrast to film's ability to capture "life," abstract, pixelated graphics offered little realism. Rudimentary game worlds, often consisting of a few lines and shapes, seemed to occupy a fundamentally different plane than celluloid. With the attainment of graphical realism some decades away, the abstract nature of arcade games positioned them as outside of traditional concepts of cultural representation. Meanwhile, conventions of narrative almost disappeared in the most ludic-based titles, such as the brick-hitting game *Breakout*. Considered largely a diversionary and fleeting pursuit, games existed within their own "cultural void." Dedicated physical spaces for video game play—the arcade and basement—as well as a lack of media attention, enforced this sense of cultural isolation.

However, on closer examination, video games filtered the same societal norms, influences, and cultural framework as other media. As Newman contends, "Contours of suburban and urban spaces, Cold War politics, transformations in labor and economics, the computerization of society, and fashions and styles in popular culture all left imprints on video games."[32] Computers simply coded things in novel, mathematical, and often unexpected ways.

Early video games digitized popular pastimes, work experiences, and recreational pursuits. A number of games recreated familiar elements of American life and sought to emulate the "American everyday." Games transformed menial, physical tasks into exciting escapades. In Apollo's *Lost Luggage* (1982), an action game programmed by Ed Salvo, the player assumed the role of a skycap tackling falling luggage from an increasingly manic airport carousel. In Atari's *Paperboy* (1985), players delivered newspapers to suburban homes, dodging hazards such as potholes, cats, skateboarding kids, and the elderly. The arcade version featured its own BMX-style handlebars to navigate with. More technical roles stretched to directing a drilling operation in *Oil's Well* (1983) and maintaining a nuclear power plant in Gottlieb's frenzied *Reactor* (1982). More abstract games similarly took their cues from everyday tasks. The idea for Atari's *Breakout* (1976), an action title based around dismantling a brick wall, came to Bushnell while he was on a beach in Hawaii, when he realized that "a game theme that tended to be successful was cleaning things up." [33]

A range of titles explored food and drink culture. Data East's *BurgerTime* (1982) challenged the player to "pick up" a list of hamburger ingredients scattered across a platform-based structure, while facing three culinary-themed enemies: Mr. HotDog, Mr. Pickle, and Mr. Egg. Mattel's *Kool-Aid Man* (1983) encouraged the satiating of creatures called "thirsties" by drinking water at a digital pool. Most famously, Midway's *Tapper* (1983) positioned the player as a bartender serving impatient drinkers, with quick reaction skills needed to keep the unquenched hordes happy as well as preventing empty tankards from falling to the floor. Graphics artist Scott Morrison, who worked on *Tapper*, explained, "The theme of sliding a glass down a bar was well established through movies and TV of the time (westerns were still very popular then), so it resonated with people and was easy to understand." [34] Technology limited how far bar scenes could be replicated; Morrison hid the movement of clientele behind the bar because "animating the walk cycles for all of the customers used way too many resources." Morrison added, "We also considered adding a digital audio chip and recorded a series of burps and belches, but decided it was too gross to keep in." With the idea to "secure the Budweiser license and sell an 'adult' video game," Midway's Tom Neiman negotiated a sponsorship deal with Anheuser-Busch. However, concern over minors consuming alcohol at the arcade led the company to recode and rebrand the title a year later as *Root Beer Tapper*.

Other titles replicated American sports. The shift from watching sports on television to playing them on a game screen seemed evolutionary and natural. As Bushnell explained, "Sports were successful because people already knew the rules, which added to their marketability." [35] The national pastime of baseball had several iterations on business computers before entering the home console market. In 1961, John Burgeson programmed the first baseball management game on an IBM 1620. In 1971, arts major Don Daglow programmed Ponoma College's PDP-10 so that players could actually play the game itself (rather than manage a team and then watch the game, akin to a television viewer). *Baseball* (1972) was one of the first games for the Magnavox Odyssey. The game had basic graphics; one team was represented by pixilated blue people, the other by red people. By the early 1980s, the national pastime was a staple on all machine formats, with versions available on the Atari VCS, Intellivision, ColecoVision, and the Nintendo NES.[36]

John G. Kemeny, based at Dartmouth College, produced *FTBALL*, the first American football computer game, in 1965 for the Dartmouth Time Sharing System (DTSS), a collection of wired computer terminals.[37] Kemeny had previously worked on the Manhattan Project and assisted Albert Einstein; his work at Dartmouth included co-authoring the first accessible computer language, called BASIC, from which *FTBALL* was programmed. An avid football fan, Kemeny wrote the program after watching an Ivy League championship game. *FTBALL* took the form of a text-based news feed; the player, as quarterback, chose from a range of seven play options, with the aim of making suitable progress down field. In the late 1970s, a surge in popularity for televised NFL games corresponded with the "golden age" of video gaming. At the arcades, *Atari Football* (1978) proved immensely popular. On a scrolling, over-head playing field, two players competed using trackballs to control their teams in short sessions (a quarter paid for ninety seconds). The same year, Atari released a home version, *Football*, for its VCS console. The game allowed only three players per team, who looked more like aliens than athletes, while the number of plays had grown to ten (evenly split between offense and defense). In 1979, Mattel launched *NFL Football* for its Intellivision console. Five-player teams competed in a relatively realistic setting, and the game boasted official NFL licensing. A reviewer for *Video* magazine gushed, "*NFL Football* will make video arcade lovers think they've died and gone to Super Bowl heaven."[38] The games *10 Yard Fight* (1983) and *Tecmo Bowl* (1987), both for the Nintendo NES console, followed. In 1984, Electronic Arts approached head coach and sports commentator John Madden to be involved in the development of a new football title. Madden insisted, "I'm not putting my name on it if it's not real." First released for personal computers in 1988, then later across console formats, *John Madden Football* was a phenomenal success. The game featured Madden himself as voice commentator and provided a vast playbook of strategies. Updated annually, the game earned a loyal following, including many NFL players.[39]

Video games replicated all kinds of American sports, from billiards to skateboarding (professional skateboarder Tony Hawk put his name on one game series). Elementary versions of basketball appeared on the Magnavox Odyssey home console in 1972 and at the arcade, thanks to

Midway, in 1974. An entertaining one-on-one version, simply called *Basketball*, followed on the Atari VCS in 1978; *Video* magazine noted that "the game definitely captures the flavor of basketball."[40] Golf games appeared on most formats, including a popular version on the NES in 1984. In Nintendo's *Golf*, the player controlled the club swing using a power/accuracy bar that became a standard of the genre. The NES similarly sported a successful wrestling title, *Pro-Wrestling* (1986), that featured realistic moves, an independent referee, and a colorful view of the ring. WWE- and WWF-sponsored titles dominated sales from the late 1980s onward. Traditional outdoor pursuits came indoors, including hunting, horse riding, and fishing. Programmed by David Crane for the Atari VCS, *Fishing Derby* (1980) depicted a side-on view of two men fishing in a lake, with players controlling fishing lines in the water. In 1997, Sega released *Sega Bass Fishing* at the arcade (later for the Dreamcast console), transforming a traditionally tranquil, leisured pursuit into a frenzied competition to catch a certain weight of fish every few minutes. An in-game commentator enthused, "Oh, that's a big one" when the player hooked a sizable catch, with rewards varying from new sporting outfits to unique lures.

These early titles, from *Pong* and *Baseball* through to *John Madden Football* and *Sega Bass Fishing*, enshrined the medium of video games as fundamentally sport-like and competitive. The opportunity to play American sports proved eminently attractive, and game consoles sold on their strength of sports titles. Video games served as an extension of fan service, part of the ritual of following favorite teams and championship events. Recognizing the power of sponsorship and endorsement, software companies courted the most famous players as well as official licenses. By the 1990s, most major gaming franchises had sponsorship deals. As video game technology progressed in the 1980s and 1990s, so too did the immersive qualities of the simulations. Programmers strived for realism and authenticity in their products. Replicating the unique feel of individual sports proved important; the sense of hitting a ball in digital tennis or virtual golf was crucial to the games' appeal. Sometimes dedicated peripherals assisted in the task, with *Sega Bass Fishing* including its own reel-able fishing rod (players felt tension in the rod as digital fish bit). In the quest for greater realism, video games captured the wider culture of American sports. Games included themed advertising, half-time shows, sports commentators, and cheerleader choreography.

More sophisticated titles such as *Madden* taught young players about the intricacies of play and even served as introductions to the game. Video games also changed the nature of American sports. They compressed lengthy competition into minutes of play and transformed complex physical actions into joystick tweaks. By focusing on spectacular moves and instant gratification, they exaggerated the entertainment and reward aspects of all sports. They transformed the fundamental meaning of sport as physical exercise into something nonphysical.

The reproduction of the American everyday in video games extended to hobbies and music. Programmers made games out of activities as diverse as bowling (Atari's *Bowling* [1979]), climbing (Taito's *Crazy Climber* [1980]), and painting (Williams' *Make Trax* [1981]). Nintendo even prototyped a knitting machine peripheral for the NES in the mid-1980s.[41]

Music took longer to synthesize in a digital form than most hobbies. While games such as *Super Mario Bros.* (1985) pioneered the use of music for game world atmosphere, few titles required players to learn instruments for themselves. In 1996, the Sony title *PaRappa the Rapper* (1996) introduced the modern music rhythm game, in which players timed button presses to individual notes, while cartoon characters danced onscreen. Two years later, *Dance Dance Revolution* (1998) required players to physically dance in time to onscreen instruction accompanied by popular music. Activision's *Guitar Hero* (2005), complete with plastic replica guitar, encouraged gamers to pretend to be rock stars by playing air guitar to classic rock tunes. Music-related peripherals grew to encompass not only guitars, but drum kits (*Rock Band* [2007]), microphones (*SingStar* [2004]), and even maracas (*Samba de Amigo* [1999]). Like with American sports, celebrity endorsement aided music-based game sales. Aerosmith, Metallica, and Van Halen granted their names to various iterations of *Guitar Hero*, and The Beatles licensed their name to *Rock Band*. Music artists lent their identities to titles outside the rhythm genre, with the rappers 50 Cent and Snoop Dogg both appearing in action titles.

Early video game depictions of America were elemental and decidedly basic. The foundations of an identifiable Gamic America came in the outline of stick figures, rectangular cars, and simple white tombstones. The first cornerstones of an America imagined in computer game form were essentially pixelated blocks. A few lines of text outlined the player's presence at digitally imagined American locations. Sometimes, only the game's title or instructions revealed the country-specific context of play. Primarily, games presented America as a space given over to sports, recreation, driving, and shooting, or at least, they encouraged the player to think that way. Game space served as an extension of American recreational space: constructions in code replicated and reaffirmed a consumer-based, play-orientated reading of the nation. Far from the game worlds of the twenty-first century, which would offer photo-realistic, fully interactive virtual cities, early video games presented America as a symbol, an outline, and a referent. Elemental codes provided the backbone to an alternate, embryonic digital realm, an electronic frontier space, full of colorful symbols, but hard to fully comprehend or navigate.

In early text-based adventure games, limited description forced players to imagine their surroundings in a similar vein to the process of reading a novel. The game space was mostly something the player mentally configured. Players worked with minimal information, inputting simple requests and awaiting computer responses. This exploratory element gave games a strong frontier aspect. The terminology of such games drew on an established language of exploration; players used directional commands and compass-based instructions, such as "go north." Gamers tackled the "unknown" with a limited range of options, often as simple as fight, explore, or hunt. The frontier appeared as a dark space (or the "fog of war") controlled by the computer, with the player cast as the torch. The full contours of the frontier could only be revealed by typing the correct sequence of commands.[42]

Text-based adventure games often drew on popular fantasy for content. The 1977 title *Zork* replicated a *Dungeons & Dragons*–style storyline and presented a world full of mythical beasts, such as orcs and dragons. Text-based titles rarely chose real-life locations to explore. However, in *Colossal Cave Adventure* (1976) for the PDP 10, programmer Will Crowther used Mammoth Cave in Kentucky as his inspiration. Widely considered the first proper adventure game and the first work of interactive fiction, *Colossal Cave Adventure* (later renamed *Adventure*) introduced the player to an elaborate and bewildering quest. Crowther (followed by Don Woods, who expanded the game) replicated the Kentucky cave in detail, transforming a physical entity into a series of lines of binary code. The game challenged players to gradually map the cave, often by using traditional pen and paper, in order to escape. Matt Barton applauded *Adventure* for its "creation of a virtual world and the means to explore it."[43]

Identifying games as set in a digital version of America often rested on the presence of simple icons, characters, or tokens on screen. Limited CPU processing and ROM memory restricted the presentation of flora and fauna. Early forms of digital nature were crude and stick-like. Simple green blocks denoted lawns, and brown sticks indicated trees. Programmers cast American fauna as protagonists; for example, the player assumed the role of an oversized but energetic amphibian in *Frogger* (1981), tasked with crossing a series of highways and rivers. Whole geographical regions could be encapsulated by the simplest of visual references: a sole cactus in the arcade title *Gun Fight* (1975) identified the digital landscape as the trans-Mississippi West. Often this binary presentation of American nature, this embryonic digital ecology, served a ludic function as well as an aesthetic one. Pixelated flora and fauna could be an obstacle, an enemy, or somewhere to hide. Nature often appeared as a barrier to the player—something to jump over, avoid, or kill—as in the classic adventure game *Pitfall!* It also defined the game space, demarcating the limits of play, as in Ultimate Play the Game's *Sabre Wulf* (1985), with vegetation marking the edges of the game's elaborate maze. Games used digital nature to set boundaries, limit movement, and challenge the skills of the player. Programmers ingeniously moved exploration of the outdoors to the indoors.[44]

American landmarks appeared in early titles mostly as backgrounds—visual interest for the player with no direct ludic purpose. A colorful Mount Rushmore appeared on the loading screen of *United States Adventure* (circa 1982) by Jerry White of First Star Software for the Apple II computer, while in *Agent USA* (1984), an edutainment title by Tom Snyder Productions, players tackled dangerous "fuzz" in cities marked by impressive skyscraper backdrops. As computer memory expanded and chip technology advanced, the complexity of the visuals increased. Colorful backgrounds became the norm for fighting games, such as *Street Fighter* (1987) and *Mortal Kombat* (1992). In the Gottlieb arcade title, *New York! New York!* (1980), players were tasked with saving the Big Apple from alien attack.[45] The gamic New York featured a pixelated Statue of Liberty in the backdrop, symbolically shining on the player as liberty's chosen protector. Similar to movies, games utilized American cities as scene setting and colorful stages where the drama unfolded. Contemporary America was one of many game stages, along with the Wild West, Medieval Rome, and the theaters of World War II. American landmarks provided titles with a welcome sense of geography, depth perception, and physicality. The visual-spatial spectacle provided a gameplay incentive; players got to see something new by completing each level.

The first view of America as a complete landmass in a computer game came in the form of a map. Included with the Magnavox Odyssey, the educational game *States* (1972), with an accompanying acetate overlay, presented a colorful map of the country on the home television. Players began their turns by twisting the Magnavox controller; a random state would light up for each player to identify. Fifty paper cards, one for each state, tested regional knowledge. The theme of education continued with *Which State Am I?* (1980) and *United States Adventure*, both for the Apple II.[46] The two games highlighted regions onscreen, and players were tasked with identifying individual states (with the additional challenge in *United States Adventure* of naming the states in order of their accession to the union). As the first way to comprehend a game space, the process of mapping situated players in the mindset of world geographers. Similar to European explorers, gamers navigated the nation as a two-dimensional cartographic creation. America existed on a single plane as a series of lines and borders. Player movement unveiled regions and territories.

Game designers dictated the form of early interaction with Gamic America, defining what players could see and what they could do. Programmers decided exactly what role each player had by codifying movement and setting pathways. Early games often worked as a series of repetitious stages based on button-pressing actions, such as running and jumping. Few titles explored the subtleties of human existence. Games made it far more normal to kill someone than to kiss someone. While in the best games players felt in control of their path, designers represented the true "architects of narrative" (to borrow Simon Egenfeldt-Nielsen's phrase) and supplied the illusion of player agency.[47] Early titles set the tone and often the boundaries of interaction and immersion; they laid the ground rules for how to navigate gamic versions of cities such as New York or defined what it meant to be an explorer on the frontier. Ian Bogost highlights the "procedural rhetoric" of games, and how they uniquely combine seeing with doing. How gamers learn to be gamers started with titles in the 1970s; people learned how to act on the digital frontier at its inception.[48]

An elemental version of Gamic America thus emerged with very different rules from its real-world inspiration. Arcade game time functioned as compressed periods of play, often measured in single minutes, with extended play gained by ludic achievement or purchased through coinage. Game space condensed and twisted notions of geography. Players could traverse the United States and Europe in minutes in the popular racing title *Out Run* (1986), designed by Yu Suzuki. Games also served the paradigm of "freedom illusion," giving the player the feeling of creating his or her destiny, but within the confines of a tightly controlled space. Early forms of control allowed players to move in four directions or follow on-the-rails systems. Early configurations of the digital frontier thus prioritized rules and restrictions. Rather than an open frontier, Gamic America closed down options and redefined progress in often linear and mathematical ways. An embryonic, BASIC-coded version of Alt-America formed around stick men and narrative rigidity.[49]

Like authors and directors, programmers chose their subject of study and their artistic take. Early digital culture homed in on certain aspects of American subject matter and turned those aspects into games. Reflective of popular culture and trends in cinema and literature, computer games picked up stories of the day. They replicated cinematic interest in science fiction and space, the old West, and the American highway. They also drew on consumer trends and sports heritage. Programmers transformed these subjects into interactive ludic realms. In the process, they often deconstructed complex subjects, reconstructing them in pixel and binary. Creative choices, market demands, notions of "play," and technological limits shaped the end product. Reflecting the language of computers, situations became logic-based, often resource-based, and eminently solvable. Commercial goals at arcades dictated that games needed to be quick (for machines to accumulate more money) and easy to access. Concerns over profit margins as well as computer memory shaped Gamic America.

Early titles emphasized competition, quick reactions, and numerical scoring. Games emphasized violent solutions to problems, ignoring issues of morality or alternative ways of thinking. Few titles openly engaged with serious social or political issues. Often deeper meanings hid behind the shell of the game, the ludic activity taking primacy over any contextual information. The arcade title *Berzerk* (1980) by Stern of Chicago, a maze game in which players avoided android attackers, clearly reflected broader anxieties over artificial intelligence. Michael Crichton's *Westworld* (1973) and Stanley Donen's visually fantastic *Saturn 3* (1980) explored similar issues on the big screen. However, in *Berzerk*, players focused on simply evading the enemy, with little time to consider the wider cultural symbols of an android nation and its repercussions. Games rarely offered direct commentary on American culture; instead, ideas lurked within the code. Unlike the scripted commentary of film or literature, games invited the player, in variable degrees, to influence the narrative. Rather than being a detached viewer, the player participated in the programmer's view on a topic, but not always in a fully obliging way. Gamers looked for cheats and quick paths to avoid the harder sections of games, circumventing the overall sense of being directed.[50]

Part of the digital frontier, the new world of gaming also assimilated a range of frontier subjects. Inventors on the technological frontier turned their attention to other pioneers. At the IBM Pavilion at the San Antonio World's Fair of 1968, one computer tested guests on their general knowledge. The terminal asked, "How about Indians? Spacemen? Geography? The Olympics?" It assumed patrons could be quizzed about a range of frontier subjects and classic sports. Games appeared caught in a narrative of replaying past achievements and re-enacting victories.[51]

Reflecting the enduring appeal of the frontier, programmers transformed the "historic frontier" and the "space frontier" into digital environments. Games became tools to experience past and future pioneer realms. Games offered unexplored places akin to blank spaces on a map and provided a "new world" for new digital colonists. They extended an imaginative frontier experience beyond the pages of James Fenimore Cooper or the charter of Apollo 11. They made the frontier itself a game. This frontier seemed distinctly limitless, endless, and free. As Edenfeldt-Nielsen et al. describe, games provided "an ever-expanding world."[52] New England colonists in the 1700s had thought of the United States in a similar way. Player experience of this ludic frontier focused on a process of systematic conquest. Driven by the idea of something new just around the corner, the reward of the next screen, players tamed the digital frontier. Processes of discovery, conquest, and victory underscored a range of early computer games, from *Adventure* (1976) to *Civilization*

(1991). The mythology of the frontier shaped the first few decades of video games.

Exploring the Oregon Trail

In 1971, a trio of students training to be teachers at Carleton College, Minnesota, created the computer game *The Oregon Trail*. Asked to teach a unit on 'Westward Immigration,' trainee history teacher Don Rawitsch "conceptualized a 'board game' representation of the 2,000-mile journey . . . But before I took that idea very far, my two college roommates (who were aspiring mathematics teachers) suggested making a computer game instead."[53] Aided by Bill Heinemann and Paul Dillenberger, Rawitsch set about transforming American history into a playable computer title. Heinemann and Dillenberger programed *The Oregon Trail* in a janitor's closet, using a teletype, a mainframe, and a printer to complete the game. On December 3, 1971, Rawitsch introduced an early version to his class. Rawitsch remembers, "They loved it. Most had never used a computer before. It was a much more active endeavor than simply reading a history book. And they were fascinated by the notion that they could learn about history from a computer activity." Welcomed by learners, Carleton College rolled out *The Oregon Trail* to other school teletypes. The code re-emerged as part of the Minnesota Educational Computing Consortium (MECC) range of programs for state schools in 1974. In the late 1970s, an improved Apple II version that sported graphics and audio became a standard educational experience for American high schoolers. Arguably the first edutainment title, *The Oregon Trail* went on to sell around sixty-five million copies, with Harry J. Brown deeming it "one of the earliest and most successful attempts to use interactive entertainment to teach history."[54]

Rawitsch sought to educate students about the American experience using novel technological means. He explained, "We approached this as a simulation model. The programming code was used to track the player's progress along the Trail, to track the player's resources that helped keep him alive, and to set the probability for each event happening to the player, along with the impact on his supplies."[55] Rawitsch transformed the Oregon Trail into an information trail. Typing commands on a teletype, the player prepared for travel. Using allocated virtual money, gamers bought a range of provisions from Matt's General Store, then, at regular stops, chose whether to hunt, trade, or explore along the trail. The title presented players with a range of data along the way: time, weather, health, food, and miles travelled, and required them to manage their resources for the expedition. The game mixed resource-focused strategy with linear adventure. It transformed the fundamentals of the trail experience into something primarily mathematical, logical, and practical. The game existed as a formula based around a series of choices. The trail functioned in essence as a logic tree. Knowing how much to spend on different resources proved crucial to success. In later iterations, players received aural rewards such as a "bang" for a success-ful hunt, and the teletype rang in celebration when players reached the final destination, Willamette Valley.

Along with the imperative of wise resource management, the game presented progress as an exercise in linear movement. It offered no re-alistic option to deviate from the route itself, with westward travel an automated path. The constant one-directional movement allowed lim-ited room for alternative actions or perspectives (Rawitsch later pon-dered how a companion game might have explored westward expan-sion from the perspective of an indigenous family). Success in the game was determined by how many miles players had travelled. Reaching Oregon (in gamic terms, winning) represented the obvious purpose of play, but Rawitsch also found that through the "twists and turns" of the adventure, players "picked up other messages as well" about planning, perseverance, and courage.[56]

A series of challenges and obstacles appeared along the route. With elements that presaged the survival-horror game genre, *Oregon Trail* focused attention on the need for party members to eat to restore their dwindling health and energy, while avoiding disease and death.[57] Anxiety revolved around what challenge would be next and what unexpected disasters lied ahead. Random events could bring death to the party. Additional challenges in the Apple II version included diseases such as typhoid and dysentery, as well as accidental broken limbs. Players faced the choice to barter or fight with Native Americans. Rawitsch taught Native American students in the classroom where he first introduced the title and "knew that the Native American issue was a delicate one in the 1970s."[58] He thus researched actual settler accounts of the trail experi-ence, "found that Native Americans were often helpful to the pioneers," and "made sure that [those] events were added to the game."

Rawitsch's *Oregon Trail* replicated some of the themes of earlier literature and film. In *Oregon Trail: Sketches of Prairie and Rocky Mountain Life* (1849), historian Francis Parkman presented the trail as an epic nation-building story.[59] Although the book only covered the first third of the adventure, it helped cement the route in popular memory. In art, Albert Bierstadt's *Emigrants Crossing the Plains* (1867) depicted the trip as one of majestic beauty and romance. Meanwhile, *American Progress* (1872) painted by John Gast cast the expedition as fundamentally about modernization and manifest destiny. The trail continued to have resonance in the twentieth century. In the Hollywood movie *Oregon Trail* (1959), director Gene Fowler employed the trail to explore issues of greater conflict, casting the Oregon Question as fundamentally a "question of war."[60] A movie scene referenced Parkman alongside (somewhat bizarrely) pickled herring, with a con artist paraphrasing the historian, "Destiny. It is written on my empty palm."

Like these earlier representations, the game captured the dominance of movement and the sense of unfolding drama. All media, including Rawitsch's computer game, fixated on the iconography of the journey, focusing on the animals, the wagons, and the people on the trail. Literature, film, and game alike celebrated the pioneer spirit and the grand narrative. However, the gamic trail diverged from early media by rationalizing the experience and by highlighting the everyday challenges of travel. The game reframed the story to be less mythic and more about resources.

The game-based *Oregon Trail* also contrasted significantly with the real frontier experience. The actual trail reframed gender roles and encouraged a new ethnic diversity in the West. The experience of the Great Migration in 1843 involved hardship, loss, and physical endurance, in equal measure. The trip challenged families, businesses, and American technology. The trail exacted huge costs. Across a 2,000-mile trip taking about six to nine months complete, an estimated 20,000 people died from diseases such as cholera. Thousands of oxen, horses, and other animals fell. Such dramatic experiences proved impossible to translate with a 1970s college computer. Playing the computer game involved no physical exertion, no exposure to harsh environmental conditions, and no threat to life. The game offered no sense of snow-enveloped mountain passages or the risk of heat stroke in the desert. The title compressed the journey into approximately twelve turns, with gameplay lasting a matter of minutes for the neophyte player. The title also featured a range of historical errors, including misrepresenting oxen as the only transport (mules, horses, and people all pulled wagons), casting a lone wagon against the elements (travelers often used a guidebook and travelled as part of a group), and depicting a trail littered with grave stones (while death occurred, grave stones were rare). The challenge of the real trail was infinitely more taxing. As one commentator wrote, "Hunting was also a lot harder—and more dangerous—than simply pressing a space bar to shoot."[61]

What *The Oregon Trail* achieved was a sense of "playful realism." Players travelled from Independence, Missouri, and, like real wagon train heads, chose when to leave and what supplies to take, and duly managed those resources. On some level, players experienced the most important elements of the frontier experience. The game recreated the sense of distance and mundane travel by its repetition of screens and focus on miles travelled. It captured a sense of peril, with wagon wheels routinely breaking, and party members suffering from a range of maladies. The title imparted a sense of adventure and traveling into the unknown. It did all this with an element of simplicity and humor—a playful realism. Taking on the persona of a Daniel Boone–type character, gamers explored a frontier mentality, but through a prism of satire and work. They played the archetype of the frontier hero, but one whom they could call silly names. The outbreak of disease on the trail became a running joke. When a party member died of dysentery, most of the classroom laughed, whereas to Francis Parkman, such a matter was "too serious a thing for a joke."[62]

The Oregon Trail subtly reframed the frontier story. Restricted by the nature of computer technology and early gameplay conventions, Rawitsch transformed the complex contours and meaning of the trail experience into something primarily mathematical and resource-based. Like a game of chess, the player visualized next moves and multiple eventualities. The Hollywood romance of the frontier was replaced with code and procedure. However, despite its numerical elements, *The Oregon Trail* successfully transformed history into a convincing personal story. Rather than read a historian's account of westward travel, players assumed authorship and control over the American past, a very novel and exciting prospect in the early 1970s. As Rawitsch explained, the team that developed the game "wanted the player to have a first-person experience. That way he can not only play a game, but he can feel history." Students thus had the opportunity to "relive history," with "an opportunity for introspection, both on worldly events and the player's personal experiences."[63] Both the novelty of the interface and the game itself proved significant. As Jon-Paul Dyson at the International Centre for the History of Electronic Games explained, "For generations of computer users, [*The Oregon Trail*] was their introduction to gaming, and to computer use itself."[64] *The Oregon Trail* simultaneously taught people about computers, the technological frontier, and the historic frontier experience.

Founding America in *Civilization*

Notions of exploration and nation-building proved popular in other early strategy titles. Doug Dyment programmed *Hamurabi* (1968), a text-based resource-management game based around being the titular Babylonian king, on the PDP-8, while Peter Langston created the war game *Empire* (1972) on a HP computer at Evergreen State College. An example of the city-building game genre, Don Daglow's *Utopia* (1981), a strategy title on the Mattel Intellivision platform, depicted two islands competing for dominance. Another early simulation title, *Fortune Builder* (1984) for the ColecoVision console, presented a space resembling the US eastern seaboard as a territory ripe for investment. By placing complementary developments close to each other (for example, a resort next to a marina, or housing next to a factory), the player grew a financial empire.[65]

In 1991, MicroProse released the frontier strategy game *Civilization*. Released on the PC, Commodore Amiga, Mac, and Atari ST, *Civilization* was a phenomenal success. Creator Sid Meier modeled his computer game on board games such as *Risk* (1957) in which players took charge of empires. Similar to the 1980 board game *Civilization*, Meier's title had a Darwinian feel with a focus on technological progress. Players coaxed their nation into action. Adam Chapman classes *Civilization* as a "conceptual simulation style" of history game, a procedure-heavy type of game designed to "*tell* us about the past without purporting to *show* it as it appeared."[66] Influenced by Stanley Kubrick's *2001: A Space Odyssey* (1968), the beginning of Meier's game depicted the evolution of the planet, the making of life, and the beginning of tribes. This sense of a new world and a new frontier had a number of ludic qualities to it.

Meier presented the player with one goal: to civilize the new world. Players chose the basic geography of that world, with the option of an abstract set of islands or a traditional world map.[67] They chose their "tribe" from a range of nationalities. Each tribe began the game with a distinct set of skills and technologies, including Americans who specialized in "democracy" and had the trait of "not being overly aggressive." By choosing the American option and a terrain that replicated Earth's cartography, *Civilization* presented the player with the opportunity to replay history and conquer America as if for the first time. Assuming "Abe Lincoln" as avatar, the player witnessed the construction of the first settlement of Washington, in game time of 3980 BC.[68]

Gameplay revolved around actions on a map. Like Meier's companion title *Colonization* (1994), *Civilization* presented game screens that resembled Age of Discovery maps. With a foregrounding of the sea, swaths of unknown territory, and simple outlines of "new land," the computer-generated pixel screens bore some similarity to hand-drawn interpretations of America as a New World in the 1600s. Focused on physical boundaries, tributaries, and settlements, early colonists drew the New World as a simple and knowable geography. There seemed little to separate *Drake's Capture of Santo Domingo* (1589), a hand-colored map engraving by Battista Boazio, that depicted Drake's search for the West Indies, from Sid Meier's game screens. New World cartography and digital cartography imagined a similar world to explore.[69]

The game version of Abe Lincoln possessed not just the town of Washington, but a wooden wagon resting on the grass. The wagon provided a convenient vehicle for exploration. By moving the wagon a square at a time, the player revealed the surrounding terrain. Game space started out as digital nothingness, a black screen, with the ludic frontier cast in darkness, as if a pixel wilderness hid heathen forms. As the wagon moved, it revealed the form of the frontier: its ecology, its resources, and its dangers. The act of movement shone light on the dark screen, bringing the promise of "enlightenment" to the frontier. The further the wagon travelled, the more the form of the country emerged. Geometric squares on the map marked the contours of the digital frontier. The process of movement identified the game as essentially "a spatial story," according to Ted Friedman.[70] Move by move, players uncovered new territory on a digital canvas.

Colonists and gamers looked for the same things in the landscape. A passenger on Walter Raleigh's Virginia trip of 1585, artist John White was instructed to "drawe to life all strange birdes beastes fishes plantes hearbes . . . the figures and shapes of men and women in their apparell as also of their manners of wepons in every place as you shall finde them differing."[71] White drew up a comprehensive inventory of the local environment. Likewise, in the game, as players moved across the squares, they revealed such resources as supplies, huts, or water. *Civilization* encouraged a view of the land as raw material to convert into something more useful. Rather than admiring the scenery, players were expected to act in a pioneering manner and adopt an expansionist attitude to the land.

The gamic Abe discovered that Washington, with a starting population of 10,000, provided a few resources of its own, including a military unit. Meier presented the first product of a civilized city as an army. The appearance of an American militia force allowed the player to not only defend the city, but also conquer the surrounding areas. Exploration led to the discovery of "barbarians and hostile armies" with which to battle. Adopting the rhetoric of manifest destiny, players wiped out competing tribes or converted them to "more advanced" loyal units. The gamic process of discovery intimately connected with the action of assimilation and conquest, facilitated by military power. Meier offered a historical picture of cultural exchange with militarism as its bedrock.

In the companion title *Colonization*, Meier expanded exchange to include trade of guns and horses, but as Rebecca Mir and Trevor Owens note, historical realism fell short of including disease as the key exchange between groups.[72] Limits to Meier's depiction of the workings of civilization also surrounded his sense of exchange as decidedly one-directional. Both *Civilization* and *Colonization* encouraged the conversion of indigenous people, but provided no mechanism for the transformation of an American into a Native. Native Americans underwent an in-game conversion process from "barbarians" to "civilized" Americans, their transformation highlighted in a visual shift from wearing animal skins to donning traditional Western garb. This assimilation of the native marked the game as a digital version of the Carlisle School mantra of "Kill the Indian, Save the Man." Fixed in computer code, the Indian had no future in *Civilization*. With the Native's fate predetermined, gamers faced no sense of moral ambiguity over this frontier decision. A restrictive game world kept the Indian in check.

The ancient city of Washington not only furnished military resources, but also provided technology and knowledge. Employing a technology tree, which simulates progress by way of stages, or branches, of research, Meier created a series of challenges for players across time, from the discovery of the alphabet through to building nuclear weaponry. Players sought the ultimate achievement of constructing the Seven Wonders of the World using their mastery of the in-game technology tree. *Civilization* offered an exciting scientific frontier to play with. Meier cast technology as linear, progressive, and ultimately utopian, similar in tone to the World's Fair white cities or Disney's Epcot Center. *Civilization* depicted life as series of technological breakthroughs, with the best kind of society one that wholly embraced science. The successful player quested after new technology with frontier zeal. From the F-15 fighter plane and Manhattan Project to the Apollo Program and Hoover Dam, American-specific examples featured highly in the game world.

Meier bequeathed the player with great powers to shape lands, fight wars, and control digital lives. The gamic frontier gave the player the notion of immense power. The player had the opportunity to "make America" and found a nation in digital climes. *Civilization* made the player feel special when computer-generated minions, feeling favorable to their leader, engaged in acts of worship, including adorning the player's palace. Looking down on events and overseeing worldwide developments, gamers assumed total control. Gerald Voorhees considers the act of play almost transcendental, and marked by a strong sense of voyeurism and looking.[73] Along with Will Wright's *The Sims* (2000) and Peter Molyneux's *Populus* (1989), *Civilization* represented a classic God game, whereby players indulged in a father archetype that bordered on megalomaniac fantasy. Similar to the players blazing the new frontier in *The Oregon Trail*, these gamers in *Civilization* acted as founding fathers, securing the fate and form of their nation. *Civilization* players took on the role of "the father figure," looking after (or sacrificing) their settlers and warriors for a greater purpose. The player, as great father, resembled how the US president was presented to real Native Americans in the nineteenth century, as a paternalistic overlord with supreme powers.[74]

In interviews, Meier claimed that his strategy titles were deliberately neutral and nonpolitical. Abiding by magic circle logic, the games were devoid of real-world meaning, confined to a virtual world of code. But both *Civilization* and *Colonization* had lots to say about society. They provided playful versions of nation-building that celebrated technology, capitalism, and modern liberal democracy. While granting players some freedom of movement, Meier consistently confined them to one model of progress. Compared to previous representations of civilization, Meier's game seemed narrowly defined.

In 1915, Thomas H. Ince produced and directed the movie *Civilization*. Ince provided an antiwar message through the actions of submarine commander Count Ferdinand. One of the earliest big-budget films, the movie cost $1 million to produce and, according to publicity, involved $18,000 in ammunition, two sunken battleships, and 40,000 horses. Ince promoted *Civilization,* both the movie and the meaning of the term, as "the love of humanity."[75] Viewers of the film took away feelings of horror about war, as well as sentiments of pacifism. In the same year, D. W. Griffith released *The Birth of a Nation*. A Civil War drama based on *The Clansman* by Thomas Dixon, Griffith's movie projected a white triumphalist story of civilized America and was duly protested by the NAACP. The first movie to be shown at the White House (Dixon was a classmate of Woodrow Wilson), *Birth of a Nation* lent racial meanings to its title.

An early example of the epic video game, Meier's *Civilization* differed from both of these film models. For Ince, civilization meant peace and humanity; for Griffith, nation-building was tied to prejudice and war. For Meier, civilization was computer led and technological. While both Ince and Griffith strived for emotional definitions, Meier idealized logic. Meier presented civilization as something mechanical, linear, and metrically quantifiable. Meier thought of civilization as a path toward greater knowledge, economic growth, victory over other cultures, and the reaching of outer space. As Voorhees notes, the concept existed as a "process" with "win" scenarios.[76] The game operated by codes biased toward certain systems of economics and politics. Players who embraced the codes of liberal democracy, capitalism, and conquest succeeded. Those who championed Western ideals triumphed.

Meier clearly modeled his civilization on US precedent and offered the American system of governance as the ideal. With the goal of transforming "empty spaces" into productive settlements, Meier's game espoused classic Adam Smith principles. Seen as a place of competition, the game world seemed Darwinistic in its preaching of survival of the fittest. Encouraged to see all land as free for the taking, players operated under the mandate of manifest destiny. The gamic Washington became John Winthrop's City upon a Hill. Whatever the outline of the digital landmass, players embarked on the familiar story of making America and followed a linear path toward the end goal of victory over other cultures. Choosing different countries or even political systems mattered little, as most of the code worked by a standard logic in which player progress flowed from mastering a liberal, capitalist game of conquest. As Christopher Douglas contended, "The game places the player in the position of guiding America's development (even if the name of the civilization we play is different); we reenact the historical-territorial drama."[77] Advertised as a chance to mold a civilization, Meier made "making America" into a game and promoted a binary version of American exceptionalism.

The game gave little space to issues and groups outside the established system, thus marginalizing indigenous people and paying lip service to such pressing issues as environmental decay. The Ameri-centric system gave few options for players to deviate from the technology tree or to replay history from a different perspective. It also romanticized and fetishized the concept of empire, tying it with the United States. As Kacper Poblocki explained, "the fetish object of Meier's fantasies is the ultimate empire, the state that resembles most the end product of all human advancement, namely the United States of America."[78] Grouping it with other fiction that historically popularized expansion, Poblocki claimed that "what *Robinson Crusoe* did for the colonial England, *Civilization* does for today's United States." However, in contrast to *Robinson Crusoe*, *Civilization* not only preached colonial ideas, but also encouraged audiences to become active participants in the unfolding drama—to carry out the conquest themselves. The game locked the player into a set of ideas with its own self-perpetuating logic. Douglas noted, "The player participates in producing an ideological effect." He also feared that "because these ideas are coded into the game rules, they appear as inevitable historical rules."[79]

Meier offered the player two paths to victory: conquer all of the nations or begin colonizing the "final frontier" of space.[80] Less warlike players won with a screen depicting the first colonies on the moon, the game ending with the announcement, "Intrepid pioneers began colonization of a new world." For Meier, territorial conquest bred success, and civilization evolved by continually annexing new frontiers. The computer model of civilization thus seemed not that different from notions of the American frontier proposed by historian Frederick Jackson Turner in the 1890s.[81] Like Turner, who saw the frontier as linear and progressive, Meier made civilization a logical process. As digital Turnerians, players explored "virgin land," built settlements, and fought nations. By constantly moving, they revealed dark spaces on the map, shifting the border between "civilization" and "savagery." They played the game from a quintessentially Turnerian code of valuing and converting resources. They explored new frontiers and beat back barbarians. Like Turner's Frontier Thesis, Meier's *Civilization* presented land and indigenous people as raw materials. The early 1990s digital frontier thus looked remarkably similar to the historical frontier one hundred years prior. Meier offered a redemptive nation-building game that legitimized, codified, and rationalized the frontier endeavor. Players worked by a Turnerian code, but without the real challenges or costs. As Douglas contended, *Civilization* continued a "national fantasy."[82]

This sense of national fantasy made *Civilization* a cathartic replay of the past. It provided a redemptive crusade. As Mary Fuller and Henry Jenkins contended, "Cultures endlessly repeat the narratives of their founding as a way of justifying their occupation of space."[83] The game, Fuller and Jenkins continued, "allows people to enact through play an older narrative that can no longer be enacted in reality—a constant struggle for possession of desirable spaces." *Civilization* thus legitimized past actions and gave America a decisive and powerful trajectory in the game that it arguably lacked in the real world.

In theory, early video games offered new possibilities to explore the American experience and encouraged players to reflect on the role of historical agency. By granting the opportunity to test historical scenarios from different perspectives, games had the potential to enhance understanding of cultural exchange, multiple frontiers, and the ambiguities of conflict. However, in the examples of both *The Oregon Trail* and *Civilization*, games mostly recycled old ideas about historical experience. They codified in binary a problematic frontier concept, then presented it to the player as the only tool for traversing a new digital domain. Rather than see the frontier as something fluid, about exchange and experimentation, games offered the American past as a linear experience about conquest and victory. The past was something to be played and won, a conflict scenario. In the process, video games reduced historical understanding. Dressing the virtual frontier as an old frontier, programmers forged a digital realm that often operated by an outdated logic. Game spaces served as escapes to a simplified world of civilization versus barbarism. In terms of historical gaming, Gamic America appeared decidedly didactic.

Looking Forward: The Frontier of Space

In terms of the frontier concept, early video games also looked forward in time and space. Aided by Kennedy's announcement of a new frontier and NASA's series of outer space missions, the cultural imagination of the 1960s and 1970s revolved around the prospects of space exploration. In many cultural quarters, interest in the old frontier gave way to excitement over the "final frontier." Originally envisaged as a celebration of the Wild West, the Seattle World's Fair was rejigged to focus on space instead.[84] At the movies, Space Westerns replaced traditional Westerns. The most watched television event for 1969, the NASA moon landing, captured a nation's imagination. In music, British artist David Bowie wrote a *Space Oddity* and soon after asked, "Is there life on Mars?"

The explosion of interest in outer space hugely affected game content. The subject spurred the imagination of technology enthusiasts and programmers alike. In the 1960s and 1970s, video games explored the theme of space exploration more than any other topic. Roughly one-third of arcade machines in the late 1970s had a space theme. Major arcade releases included *Computer Space*, *Space Invaders*, *Asteroids*, *Galaxians*, *Lunar Lander*, and *Berzerk*.

Programmers sought to replicate the process of space exploration and the machinations of the space race. In popular culture, the Apollo 11 Moon Landing served as a visual spectacle, and a symbol of national triumph. Courtesy of live television coverage, millions watched the lunar module Eagle land on the moon, with astronaut Neil Armstrong narrating, "One small step for man, one giant leap for mankind." The same year as the landing, Jim Storer, a seventeen-year-old Massachusetts high school student, encouraged by a teacher who taught programming and assisted by his father, an electrical engineer, developed the first game based on the Apollo mission as a school project, working with a PDP-8 and Foxal language. In an interview, Storer noted, "The Apollo moon landing was a huge event at the time and also lent itself well to a simple text-based program that could be implemented within the limits of that system . . . I remember that I wanted the code to be as realistic as possible."[85] Titled the *Lunar Lander Game*, Storer's version of the Apollo moon landing translated the NASA mission into a series of mathematical equations. Rather than gasp on seeing a new planetary surface, the player watched as lines of data filled the screen. *Lunar Lander* focused on calculations to determine the correct burn rate of space fuel and appropriate velocity as the Eagle headed to the moon. As if remotely controlling the enterprise from a NASA control room, players exercised distant control of the process. The otherwise blank screen generated a sense of space as alien and vacant. Using just fifty lines of code, the game promoted comprehension of the event as an output of a numerical formula. It codified the moon landing as a carefully orchestrated mathematics test. Storer "let friends play the game and . . . submitted the program to the Digital Equipment Computer Users Society (DECUS)." *Lunar Lander* became a popular game for people to play on their own computers at home in the 1970s.

In 1973, DEC business computers commissioned Jack Burness to program a graphical simulation of the Apollo moon landing to demonstrate its new DEC GT40 graphics terminal. Burness visited MIT, where the real lunar lander had been co-designed. Called *Moonlander*, his simulation used vector graphics and was featured at several DEC national trade shows. When it was later converted to the RT11 computer, Burness' game went through subtle modifications and re-codes. Fantasy and satire entered the game world, with players discovering a McDonald's restaurant hidden in the lunar landscape. If successfully piloted to the right coordinates, an astronaut could be seen exiting the lander and walking directly to the fast food outlet. In 1979, Atari released an arcade version of the game featuring similar vector graphics. Players pressed a button to apply thrust and used a joystick to carefully rotate the spacecraft. The title put action ahead of realism. Defending the shift to a simpler take on the moon landing, Atari engineer Howard Delman explained, "Not everyone is trained to land a spacecraft on the moon."[86]

Early video game content reflected a mass interest in mapping the vast unknown. The sense of space as an unexplored territory inspired one of the most successful science fiction franchises, Gene Roddenberry's *Star Trek*, first shown on television between 1965 and 1968. As Captain Kirk's monologue at the beginning of every episode related, "Space, the final frontier. These are the voyages of the Starship *Enterprise*." America was interested in the next stage of frontierism. A liberal utopianist, Roddenberry projected a vision of the future in which currency, racial conflict, and poverty had all disappeared. *Star Trek* itself became a popular subject for games. Programmer Mike Mayfield provided the first iteration, written in 1971 on the SDS Sigma 7 Mainframe, a commercial 32-bit machine. Mayfield transformed the television series into a numerical and textual game. Televisual action morphed into typing keywords on screen, such as "Kill Klingons." The *Star Trek* frontier became a series of options: report, starboard sensors, galaxy map, phaser, torpedo, warp engine. The frontier was, to paraphrase a well-known Vulcan, highly logical.[87]

Game designers conceptualized the space frontier as a game of simple conflict and conquest, not that different from Turner's take on the old frontier. Titles based around dogfights were commonplace. *Spacewar!* (1962), *Space Travel* (1969), and *Computer Space* (1971) all presented the frontier as a duel. The theme of space conflict was not reserved to US climes. Japan-based Taito's *Space Invaders* pitted the player as a lone hero against a formidable invasion force of alien vessels. Invaders made their way down the screen in regimented fashion, firing at regular intervals, while the player, at the bottom of the screen, hid behind barriers as well as shooting back. *Space Invaders* presented the aliens as the clear aggressors. It inverted the frontier thesis, to be not about discovery, but the defense of home. The game became a huge hit across American arcades, selling 60,000 units in 1979 alone.

While *Space Invaders* projected the frontier as a line of danger moving downward toward the player, *Defender* (1981) by Williams employed parallax scrolling and depicted a frontier space that continually refreshed. Other titles such as Namco's *Galaga* (1981) or Atari's *Tempest* (1981) utilized a series of progressively more difficult frontier spaces, as screens filled with speedier or more aggressive aliens. Space games came to present a frontier far more expansive and open than either Turner's frontier or the linear path set for players in *The Oregon Trail* and *Civilization*. Futuristic titles showed an important reality about the frontier: it lacked a set direction. While space games were spatially and aesthetically inventive, the traditional theme of conquest nonetheless persisted in the genre. Exploring the "final frontier" still seemed mostly a task of shooting things. Akin to a mythic Custer's Last Stand in space, players faced endless threats. The new frontier involved nonstop killing of enemies. Like the Space Westerns of *Star Wars* and *Outland*, gamic space resembled high noon with laser pistols, as players were caught in a perpetual shootout.[88]

Digital Frontiers

Featured at the New York World's Fair in 1964, the Westinghouse Time Capsule II aimed to provide a document of world "history, faiths, arts, sciences and customs," to be opened 5,000 years on. It contained forty-five items in four categories. Cast as the two key breakthroughs of the twentieth century, space and atomic energy were each a distinct section, with everything else listed under common use or scientific developments. The time capsule included a heat shield from the Mercury Aurora 7 spacecraft and a carbon-14 radioactive isotope, along with cigarettes, a Beatles vinyl record, a Polaroid camera, and the bible. Under scientific developments, Westinghouse included a single UNIVAC computer memory unit. If the oracle was to be believed, the atomic age and space age would fundamentally reshape the world ahead and determine the path of humanity. Items such as the computer chip would play a minor supporting role in a much bigger drama.[89]

By the 1990s, it was clear that the Time Capsule I from the 1930s had made some poor predictions about which items would shape the twentieth century. The first capsule contained such defunct and irrelevant items as slide rules, microfilm, and linotyping that had stayed in use barely fifty years after the time capsule was sealed. Time Capsule II had also miscalculated. Westinghouse scientists underestimated the effect of the UNIVAC chip on American life. Thirty years on from the New York World's Fair, Americans could click on the World Wide Web, text using cell phones, play computer games at home, and even try early forms of virtual reality. The real frontier of America in the late twentieth and early twenty-first centuries was digital, not lunar or atomic. Outwardly the New York World's Fair sported space-age Googie architecture, but the real architecture of the future would be Google.

In 1997, scholars Mary Fuller and Henry Jenkins commented on the meaning of the new computer-based frontier. They compared Nintendo's platform game *Super Mario World* (1990) to narratives of the New World, and noted the similarity of "the physical space navigated, mapped, and mastered by European voyagers and travelers in the 16th and 17th centuries and the fictional, digitally projected space traversed, mapped, and mastered by players of Nintendo video games."[90] Players operated by the same logic as historical pioneers. They played *Mario* out of an appetite for discovering and appropriating new spaces. "What never loses its interest is the promise of moving into the next space, of mastering these worlds and making them your own playground," remarked Fuller and Jenkins. Games enabled the "recycling of the myth of the American New World," with Nintendo transforming players "into virtual colonists." The act of gaming extended America into new imaginative realms, of virtuality and simulation, powered by a time-honored frontier modus operandi.

The programmable frontier also promised a perfect fantasy for its explorers, free from any responsibility or risk. Fuller and Jenkins suggested, "Part of the drive behind the rhetoric of virtual reality as a New World or new frontier is the desire to recreate the Renaissance encounter with America without guilt."[91] Online worlds and computer games provided fundamentally safer spaces for the frontier endeavor. Similar to Disney theme parks, games offered anodyne and cleansed spaces that bypassed the 'nasty' stuff of slavery, internment and racial killing, but crucially serviced feelings of personal success allied with national agendas. As Italian-American plumbers, virtual soldiers, and digital pioneers, players embraced the frontier spirit.

The intersections of frontier, play, and computer technology promoted at the IBM Pavilion in the 1960s thus endured. The act of "frontiering" provided a core activity in early gameplay. Gamers acted akin to New World explorers. They navigated a new geography and led a digital conquest.

2

Playing Cowboys and Indians in the Digital Wild West

In 1982, Walt Disney Studios released the science fiction movie *Tron* at the cinema. Written and directed by Steven Lisberger, *Tron* related the story of Kevin Flynn (played by Jeff Bridges), a maverick computer programmer who "downloads" himself into a dangerous computer game world powered by a corporate mainframe. Inspired by the arcade machine *Pong*, Lisberger's movie explored the new frontier of computer gaming. Featuring the latest in computer-generated animation, *Tron* depicted Flynn riding cyber-cycles across a distinctive, vector-graphic world. Dazzled by the special effects and futuristic plot, Roger Ebert from the *Chicago Sun-Times* remarked, "This is an almost wholly technological movie." *Tron* reflected a surge in popular interest in computer technology. Video games had finally entered mainstream culture.[1]

The same year as the release of *Tron*, US brand Atari had its most successful year in the video game business. Along with major success with arcade machines such as *Pong*, *Tank* (1974), and *Asteroids* (1979), Atari reaped the rewards of its first cartridge-based home console, the Atari VCS (later known as the Atari 2600). Introduced in September 1977, the console, with its wood veneer that mimicked home television sets, quickly became a household item. Although it was not the first game system for the home, the VCS became popular due to its flexible programming base, effective advertising, wide range of titles, and easy-to-use game cartridge system. With the VCS, Atari successfully brought gaming into the home.

The rise of the home machine was matched by growth in the arcade sector, where the number of establishments had grown to around 13,000 by 1982, marking a national high.[2] On 30 November 1982 in Ottumwa, Iowa, Mayor Jerry Parker held a parade to celebrate the town's burgeoning video game scene and its hosting of the first North American Video Game Olympics. He declared Ottumwa to be the Video Game Capitol of the World. Arcade owner Walter Day later called the town the Dodge City of video games.[3]

Video games also spread to shopping malls, with 221 Aladdin's Castle arcade zones, owned by pinball manufacturer Bally, operating by 1981. Families saw game titles at the local mall and arcade that they could then purchase to play at home. Popular arcade conversions included Taito's *Space Invaders* and Atari's *Breakout*. By 1982, the Atari VCS featured a catalogue of over 300 official titles, and the company boasted an 80 percent share of the American domestic gaming market. Game Studies scholar Simon Egenfeldt-Nielsen calculates that around a quarter of American homes had a game console in the period.[4] Recalling Atari consoles selling in droves throughout 1982, programmer Chris Crawford remarked, "It's difficult to convey the wild gold rush feeling that pervaded the industry that year."[5]

Then, all of a sudden, the American video game bubble burst. The home software market collapsed in 1983 due to an unexpected fall in consumer interest. Poor business practices, market saturation, widespread overstocking, the rise of home computers such as the Commodore 64 and Atari 400/800, and a series of poor-quality arcade and film conversions, particularly *Pac-Man* (1982) and Steven Spielberg's *E.T.: The Extra-Terrestrial* (1982), fueled the demise. With share prices plummeting, Warner quickly sold off its stakes in Atari, while industry innovation shifted to Nintendo and Sega in Japan.

Atari personnel allegedly buried around 700,000 unsold VCS cartridges in an Alamogordo landfill in New Mexico, close to the Trinity nuclear test site. Atari spokesperson Bruce Enten deflected local media attention by describing the truckloads heading to the city landfill as full of "by-and-large inoperable stuff."[6] However, rumors spread of a mysterious "great mass burial" of software. The western desert, which in the popular imagination is often a wasteland of dried bison bones, Indian reservations, ghost towns, and nuclear test sites, had once again become a dumping ground: this time for the Silicon Age. The dumping of unwanted *ET* cartridges in the American West became a popular cultural symbol of the rise and fall of the early video game industry.[7]

By the time of the crash, video game programmers had begun to offer their own tentative takes on American stories, in particular the legend of the Wild West. As we saw in chapter one, the wild frontier depicted in Rawitsch's *The Oregon Trail* appealed to a mass audience. Alongside space encounters and puzzle games, software companies based in both the United States and Japan drew on the nineteenth-century American West for the settings and stories in their games. Programmers transformed this history into interactive digital play and granted the cowboy a new pixelated appearance. In the process, video games inevitably added their own take on American culture, story, and legend. Games designers made their own myths of the West.

A number of Wild West titles emerged at the outset of the video game era, in the 1970s and 1980s. These games pioneered new play mediums and new technology, while introducing many Americans to digital play for the first time. Released in 1975, Midway's *Gun Fight* featured the first microprocessor, the Intel 8080, inside an arcade machine, and helped establish the shooter genre in gaming. Western games courted early public controversy surrounding video game morality; Mystique's adult title *Custer's Revenge* (1982) became the target of protest and condemnation for its portrayal of sexual violence toward Native Americans. While *The Oregon Trail* introduced schoolchildren to frontier history, action games based around the Wild West encouraged people to "play the past" at home and in the arcade for the first time. In playing Western-themed titles, Americans were transported to a mythic era of Colts and cacti, shootouts and sagebrush. They entered an interactive fantasy about historic America.

Simulating the Wild West

As American arcades in the 1970s expanded to feature electronic games alongside mechanical pinball machines, the quest began to attract clientele to this new form of entertainment. A number of titles employed popular motifs from the old American West to rouse consumer interest in the new medium. Alongside science fiction titles like *Space Invaders* and *Asteroids*, arcade machines such as *Gun Fight* and *Boot Hill* were a staple of arcade entertainment. Western-themed arcade machines offered an ephemeral escape to a "simpler" world of six-shooters, wagon trails, and iron horses. Advertisements for Midway's *Boot Hill* resembled wanted posters. However, rather than try to replicate life in the trans-Mississippi with any accuracy, these games relied on a broader mythology and culture surrounding the Wild West. Games regurgitated old Western myths and cheap novella fiction. The allegorical West—as presented by artists such as Frederic Remington, fiction writers such as Zane Grey, and Hollywood directors such as John Ford—was fed into the modern computer and spat out in pixilated form. In essence, the digital West started out as simulacra of simulacra.

Central to the recreation of nineteenth-century life in a digital domain was the reanimation of classic heroes and villains. Setting out to capture a deeply mythological West via the computer chip, game designers in the 1970s and 1980s re-animated famous figures of Western folklore as game characters. The Exidy shooting game *Cheyenne* (1984) featured a roster of legendary Wild West heroes alongside jocular, pretend villains. The fictitious, burlesque Petticoat Floozies of Lotta Love and Elvira rubbed shoulders with avatars of the real-life Dalton Gang, while tomahawk-throwing Apache Braves included Geronimo and Sitting Bull as well as the made-up Running Nose and Buffalo Breath. Characters abided by cultural stereotypes popularized in nineteenth-century Western art, in dime novels, and in twentieth-century Hollywood films. Digital women behaved as petticoat lovers, gartered prostitutes, or sassy gunfighters in the image of Calamity Jane.[8] Overwhelmingly, the dominant male protagonists were white cowboys, with Mexicans relegated to the role of corrupt banditos and African-Americans absent.

Sometimes the player took on the persona of a generic stick-figure cowboy, a "man with no name," as if lifted directly from a Sergio Leone movie. At other times, video games featured cowboys and gunslingers of significant historical repute. Atari's arcade game *Outlaw* (1976) gave the player the option of being Billy the Kid. Buffalo Bill appeared as an adept, gun-slinging nemesis of Sheriff Quickdraw in the computer game *Gun Fright* (Ultimate, 1986), the two men battling for patriarchal control over the imaginary frontier town of Black Rock. It mattered little that one character was fictional and the other based on a historic figure. As long as the gaming environment resembled public expectations of the Wild West, and thus included its fair share of cowboys, outlaws, and Indians, the dubious authenticity of the simulation could be overlooked.

The presence of Buffalo Bill as a character in a digital Western fantasy seemed eminently just. The real William Frederick Cody had merged fact and fiction in his celebrated Wild West live shows at the end of the nineteenth century. Cody recognized in the West a raw excitement that was absent from the "civilized" American East and European countries. Blending explosive action with imperialist nostalgia, Cody recreated, with consummate artistic license, mail coach ambushes, train rides through unfriendly Indian country, and historic events such as the Battle of the Little Bighorn. His theatrical productions traveled the world, underlining the American West as an experience readily consumed by a global audience. The shows were spectacular, varied, and, above all, action-packed.[9] As Thomas Altherr noted, "Most customers must have left the arena seats satisfied, having witnessed a full variety of western Americana."[10] Despite his own involvement in various frontier enterprises and a benevolent attitude toward Native Americans behind the scenes, Cody's interpretation of Western history had remarkable flaws. Cody presented the bullet as the maker of the West, sidelining the contributions of the plough and the dollar to Euro-American progress. As noted by historian Richard White, Cody popularized a highly questionable narrative of inverted conquest in his shows.[11] Always the victims of Indian aggression, Cody's white cowboys suffered few pangs of guilt as they blasted their way out of trouble.

This conscience-free conception of Western life filtered into arcade games. Games were digital Wild West shows. In Boston, New York, and San Francisco—where Wild West stage shows had once entertained crowds, arcade game machines staged elaborate and familiar fights between "whooping Injuns" and all-conquering cowboys. Programmed by Kevin Blake, *Buffalo Bill's Rodeo Games* (1989) even replicated the format of Cody's shows, inviting the player to take part in six events that tested a range of frontier skills, including knife throwing and stagecoach rescue.[12] Cover art for *Rodeo Games* heavily mimicked original show posters from the late nineteenth century. Titles such as Tad's *Blood Bros.* (1990) encouraged players to casually massacre digital Indians without rebuke or consequence. With his feather bonnet, war dance, and tepee, the pixilated Native American resembled the classic Plains Indian stereotype captured in Cody's traveling extravaganza.

Like Cody, game companies recognized the global marketability of gunfighter glory and presented the American West as an exciting and exotic mélange of six-shooters, wild horses, cowboys, and Indians. Both entertainment genres offered participants the opportunity to experience high-noon shenanigans and feel part of the historic West. Thus, the arcade machine emerged as the late-twentieth-century "pardner" to Buffalo Bill's Wild West.

Games also resembled interactive dime novels. Cody was both a product of and a contributor to the dime novel craze of the late nineteenth century. Dime novels helped cement a popular view of the West as a place of entertainment. Starting with *Buffalo Bill, the King of the Border Men* (1869) by Ned Buntline (a pseudonym for army veteran Edward Zane Carroll Judson), Cody became the subject of hundreds of dime novels. Buffalo Bill himself authored select titles, such as *Death Trailer, the Chief of the Scouts, or Life and Love in a Frontier Fort* (1878). Video game Westerns drew on the dime novel presentation of the West as rich in adventure, but bereft of realism. Mass-produced 25-cent arcade machines served as interactive dime novels of the 1970s and 1980s by bestowing affordable and accessible figments of Western reverie to all classes. Both the arcade game and the cheap novella provided simplistic, sensational stories of lovable outlaws and Indian peril, heavily steeped in caricature but light in originality. Generic, indistinguishable characters in dime novels took aesthetic form in computer games as nameless pixilated cowboys. Shrewdly identified by Henry Nash Smith, the "objectified mass dream" of Western adventure offered within every dime novel had a significant audience.[13] Such a dream later transferred easily to the computer screen.

As a predominantly visual medium, the Western video game also took its cue from the Hollywood Western. Successive Hollywood (and, later, Italian) Westerns exported the American West on a scale that William Frederick Cody could only have imagined. Cody himself was an early convert to the new technology of film, producing and starring in *The Life of Buffalo Bill (Col. Cody)*, released in 1912.[14] The amusement industry built on Hollywood's success with a range of Western-themed pinball and shooting titles. The electromechanical arcade title *Wild Gunman* (1974), by Nintendo's Gunpei Yokoi, connected a 16-mm projection screen to a lightgun, with the film reel (and story) progressing only if the player fired accurately when a gunslinger's eyes flashed. Intriguingly, the Western video game arrived during a lull in the Western movie franchise. Released in 1976, *The Outlaw Josey Wales* coincided with the debut of Midway's shooting game *Gun Fight*. The gamic West then began where *The Outlaw Josey Wales* left off. Gun fights and sagebrush, heroes and anti-heroes shifted mediums, from one screen to another. The global reach of the mythology of cowboys and Indians allowed Japanese and American game companies to produce an array of Western titles that could be easily digested by players. Thus, the familiar format of the celluloid Western enabled the new digital West to be realized.

Programmers exploited a populist, international understanding of US history and iconography based around Hollywood. They simulated an already simulated West that was comprehended by all. The arcade Western built on the reputation of its luminary forebears in the entertainment industry. Game worlds imitated film sets, with computer characters walking into view before delivering precoded scripts and actions. Game designers followed filmic conventions laid down by such Westerns as *Stagecoach* (1939) and *High Noon* (1952) in presenting familiar scenes of ambushes, chases, and grand finale shoot-outs. Players entered barroom brawls straight out of the movie *Shane* (1953). Western shooter games, such as *Gun Fight* and *Blood Bros.*, enabled players to become involved in "winning the West" in a way that cinema audiences had long desired. The audience finally had an active role to play: victory over the West was theirs to be had. Games borrowed heavily from cinematic Old West nostalgia and Hollywood's partisan exploration of How the West Was Won. The good–evil dichotomy rife in John Wayne movies reemerged in game lore. As in motion pictures, video games rarely explored the middle ground between victory and loss. The simulations were simple in their politics.

What the Wild West show, dime novel, and Hollywood Western had portrayed, the video game reproduced. Games followed the rules laid down by prior simulations of the frontier West, drawing on the same archetypes and good-versus-evil narratives. Nineteenth-century dime novels portrayed the frontier as a romantic realm of action, while movie directors transformed the Western experience into musicals, such as *Paint Your Wagon* (1969), and comedies, including *Blazing Saddles* (1973). The video game equally cast the historic West as primarily a realm of lively entertainment. The core purpose of early simulation emerged as interactive amusement, not historical realism. A natural extension of the imagined world formulated in Hollywood films and dime novels, the arcade West deviated little from an established script. Players followed the rules on screen, participating in a familiar drama. Gamers played as cowboys, beat off Indians, and conquered the American West once again.[15]

Violence on the Electronic Frontier

Given the infancy of computer technology throughout the 1970s and 1980s, games designers were restricted in how far they could represent their "historic" West. Meager computer memory and poor graphical capabilities compromised any initial plans for a comprehensive vision. Plot details, overarching stories, and character depth were all sacrificed in order to limit computer memory expenditure. Any sense of narrative gave way to an overwhelming predilection for action. Action in the game almost always meant hostile exchange. Early titles depicted duels, heists, shootouts, and ambushes. The Wild West of gamic America emerged as a place structured and given meaning by violence. The "shoot-em-up" content reflected the "pick up and play" notion of 1970s gaming—after all, few people wanted to spend hours in a dimly lit arcade reading text. Titles encouraged quick response times, with early video games fostering trigger (or joystick) itch among their fanbase.

Action also meant that the digital West differed in one crucial way from previous canvas, paper, and celluloid Wests. Unlike dime novels and Hollywood Westerns, in which immersion derived mostly from consumers' imagination and observation, the digital West allowed physical interaction from its visitors. Although Western movies inspired children to play cowboys and Indians in school yards, with the video game, the audience became truly active participants in the unfolding drama. In this way, the video game West resembled the theme park more than the movie house. Like navigating an amusement park, players in arcades entered a themed stage. At Frontier Village Amusement Park, in San Jose, California, open in the 1960s and 1970s, visitors partook in burro, canoe, and miniature railroad rides; they could also board a rollercoaster dubbed the Apache Whirlwind. More famously, Disney's Frontierland in Anaheim, California, offered stagecoach and mule rides, an Indian village, and the Mark Twain riverboat passing by a burning settlers cabin.[16] Video games replicated the same kind of action-packed thrills. However, in stark contrast to Disney's family-friendly take on the trans-Mississippi, where happy endings dominated, the arcade Western always highlighted violent confrontation in the frontier endeavor, and it encouraged the player to think and act likewise.

Cody's showmanship and Clint Eastwood's deadly accuracy rematerialized in the first Western-themed arcade title, Midway's *Gun Fight*. The game presented the Wild West in the simplest of guises. Dave Nutting produced a game based around a single deadly duel. Sporting a dark wood grain case appropriate to old saloon furniture, the machine hid its technological roots in the neon-lit arcades. Its closest rival, and influence, was the Western-themed shooting gallery, a staple of seaside resorts and fairgrounds, where players queued to take shots at ducks and other targets, sometimes for token prizes. A poster for *Gun Fight*, with two hand-drawn cowboys in the midst of a shootout, resembled a dime novel front cover in its promotion of lively action and deadly confrontation.[17] The game mechanics of *Gun Fight* resembled Atari's coin-operated table-tennis game *Pong*, with players maneuvering their characters at opposite ends of the screen while watching bullet-shaped pixels ricochet off the screen's "walls." Following instructions to "kill your opponent before he kills you," arcade customers faced off against each other on a monochrome screen that depicted white cowboys, several thorny cacti, and fast-moving wagon trains. Rather than memorizing complex instructions or game mechanics, players drew on ingrained notions of cowboy action, knowing instinctively how to act. Two pistol handles took the place of joysticks, pioneering the use of replica lightguns in games. The gamic West was a realm of violence without explanation, without context, and almost without meaning. Players shot each other for points. *Gun Fight* established a template for one of the oldest genres of digital entertainment, the Western shooter, as well as the shooter video game in general. Arcades became spaces dedicated to duels, gunfights, and digital displays of violence, with players regularly performing gunfighter roles.

In 1977, *Boot Hill*, also a Midway title, emerged as a direct successor to *Gun Fight*. Named after the first cemetery at Dodge City in the 1870s, *Boot Hill* encouraged players to imagine themselves as gunslingers entering the most violent town in the West. Although historians such as Robert Dykstra have proven that homicides were, in reality, rare in the infamous cattle town, arcade players recognized only the mythic Dodge City of perpetual violence.[18] *Boot Hill* took its cue from the scabrous boomtown portrayed in nineteenth-century newspaper reports; Stuart Lake's biography, *Wyatt Earp: Frontier Marshall* (1931); and 1950s television Westerns such as *Gunsmoke*. The game drew on the mythology of the town and its core identity, which was based on violence and bloodshed. Gamic America emerged as mythically violent.

Midway highlighted the essence of Dodge City as a series of explosive gunfights. A computer-controlled cowboy, capable of outgunning most arcade newcomers, patrolled the screen, reminiscent of Yul Brynner's ruthless android gunslinger in the movie *Westworld* (1973). *Boot Hill* paralleled robots running amok in the frontier theme park of Westworld by employing a computer character similarly disposed to impersonating and killing its human counterparts. Pixilated cowboys, identical to those featured in *Gun Fight*, uttered "Bam. Shot Me" when hit, their bodies rising up to the heavens (or the top of the screen), before turning into pure white tombstones. In keeping with the folklore of Dodge City's quick-draw duels, dead game characters kept their boots on while en route to the cemetery. A somber computer recital of Chopin's *Death March* accompanied each trip to the Hill. With only a wagon and a cactus between two trigger-happy contestants, on-screen encounters were notoriously short-lived. Arcade players competed for the highest kill tally, their anonymous victims identified as simple numbers at the top of the screen. Not surprisingly, five-minute sessions on *Boot Hill* exceeded the total body count for Dodge City over its entire history.

While representing only a small percentage of arcade titles on offer at the time, the 1970s arcade Western, through its popularity, helped cement an expectation of violence in electronic gaming. Reflecting their literary and filmic roots, Western shooters relied on the myth of the lone gunfighter as their pivotal narrative. Recognizing their target audience as teen and twenty-something males, electronics companies recreated the classic outlaw with attitude, providing digital imitations of John Wayne in *Stagecoach* (1939) and Clint Eastwood in *A Fistful of Dollars* (1964). With a few dimes in their pockets, gamers indulged their Western fantasies, playing renegade cowboys inside city arcades in palatable, five-minute doses. Video games transformed the stereotypical computer geek into the archetypal rugged individualist, the young player into Billy the Kid. Prolonged game play fostered the go-it-alone mentality of the mythical solo gunfighter—a mentality conducive to an entertainment medium focused on the performance of the individual player.[19] By inserting a coin into the Western shooter, players purchased the opportunity to see themselves as the last good cowboy in a whirlwind onscreen adventure. The white cowboy emerged as an archetype video game hero. Everyone wanted to play him. Like Clint Eastwood's Man with No Name, players became pixel cowboys with no clear background or biography. The digital cowboys simply moved and shot people. They saved damsels and maimed killers. The appeal of being among the last of an honorable breed of masculine heroes showed that the West retained currency as a redemptive crusade. The video game perpetuated a mass illusion.

Customary paperback and film themes of law and order, frontier justice, and the combating of violence with violence similarly infused the arcade Western. Prior to the advent of the electronic shooting game, dime novels, Wild West shows, and Hollywood Westerns (such as *The Wild Bunch,* 1969) cast the American West as a mythic landscape marked by confrontation and killing. As Richard Slotkin explains, tied up with regeneration, nationhood, and individualism, violence had always played its part in frontier psychology.[20] Conditioned to think of cowboys as gunmen and Indians as warmongers, arcade clientele expected games set in the West to offer confrontation. Most American and Japanese electronics companies responded with raucous, action-based titles that presented the nineteenth-century West as one continual showdown. Success in Sigma's *Wanted* (1984) rested on the quickest possible annihilation of every human protagonist; the length of each game level was determined by a preset kill tally. Capcom's *Gun.Smoke* (1985) provided an overhead view of an unfriendly town, with the player controlling a stocky outlaw on his stroll through dusty streets, dispatching all and sundry with his trusty Winchester rifle. *Express Raider* (1986) by Data East transposed the violence to the roof of a moving train. Players sparred with (rather than shot) a motley crew of bandits, while punching inquisitive coyotes for extra points. Mimicking prior fictionalized gun-blazing Wild West shootouts, game scenarios typically granted players little time to think, requiring them to rely on instinctive trigger fingers.

Two of the most violent titles were *Blood Bros.* and Konami's *Sunset Riders* (1991). The first level of *Blood Bros.* featured tin-can targets, wooden saloons, and well-dressed ladies, who responded to gunfire by lifting their petticoats.[21] Reminiscent of false-front Hollywood movie sets, building facades collapsed to reveal monumental desert landscapes. A later level consisted of massacring an Indian village, while another resembled Leatherstocking's clash with the Hurons at Glens Falls in James Fenimore Cooper's *The Last of the Mohicans* (1826). Hogs and horses, when shot, rewarded players with dynamite, guns, and bonus points, endorsing Richard White's theory of real-life Western fauna as "animals of enterprise" or "biological dollars."[22] From runaway trains to kamikaze cavalry squads, everything and everybody represented a legitimate target. The Old West appeared decidedly self-destructive and amoral.

The scrolling action game *Sunset Riders* continued in a similar vein. Players moved from left to right, jumping to avoid cattle stampedes (with the screen "shaking" under their combined weight), retrieving liquor and damsels from saloons, and challenging wanted criminals who delivered digitized speeches. *Riders* relied on the Italian Western for inspiration, introducing its major game characters—Steve, Billy, Bob, and Cormano—by way of a movie reel. In a poignant scene at the end of the reel, the four gunmen rode off into the sunset. As in the case of the classic Italian Western, few people were left alive in their wake.

However, the most controversial Western video game dealt not with guns, but with sexual license. A puerile attempt at titillation, Mystique's *Custer's Revenge* (1982), released on the Atari VCS console, reincarnated George Armstrong Custer as a computer game character seeking retribution against the Sioux nation for his defeat at the Battle of the Little Bighorn. Assuming the role of Lieutenant Colonel Custer, players maneuvered a large man, dressed only in a blue cavalry hat, scarf, and boots, across the screen, avoiding falling arrows, before raping a naked squaw tethered to a pole outside her smoking tepee. Rhythmic Native American drum pounding preceded a climatic cavalry bugle tune.

Part of a new range of adult-only titles, *Custer's Revenge* presented a misguided attempt to broaden the market of video games in the early 1980s.[23] Keen to maintain its family image, Atari responded with a lawsuit against developer Mystique and manufacturer American Multiple Industries, citing unauthorized use of Atari trademarks and negative association. Atari President Michael Moone stated, "Atari, like the general public, is outraged by this conduct"; he emphasized that the VCS was a machine designed purely for wholesome family entertainment. Women's groups, including Women Against Pornography and the National Organization of Women, and Native American organizations, including the American Indian Movement, came together to protest the release of *Custer's Revenge*. In New York, a hundred-strong crowd gathered to demonstrate outside a technology show featuring Mystique's new titles. New Yorker Kristen Reilly from Women Against Pornography denounced the software company for blatantly promoting "attack and rape," while Michael Bush, based at the American Indian Community House, interpreted *Custer's Revenge* as "a reinforcement of the stereotyping of American Indians as something less than human." In Oklahoma City, the local Native American Center picketed the local adult bookstore. Native representative Frances Wise explained, "We buried Custer once in 1876 and now we have to beat him again," while Casey Camp-Horinek chanted, "Rape is rape." The store manager returned his latest shipment of twenty-four copies reticently—twelve of them had already been preordered by customers, stating "I don't think the game was any worse than the ones children play at the arcade." Most stores refused to stock the title. State houses of representatives, including that of Oklahoma, considered adding video games to obscenity laws.

American Multiple Industries adamantly defended the title. President Stuart Kesten explained, "Our object is not to arouse, our object is to entertain." He added, "When people play our games, we want them smiling, we want them laughing." The creator of *Custer's Revenge*, Joel Miller, insisted that the title did not in any way condone rape and tenuously claimed, "He's seducing her, but she's a willing participant." The ensuing controversy boosted sales figures to some 80,000 units, roughly double that of Mystique's other soft pornography titles.[24]

Custer's Revenge shocked Americans with its graphical (as opposed to graphic) depictions of sexual violence. The title also diverged from traditional portraits of the Westerly experience, in particular its perversion of the last stand mythology. While slaughtering Native Americans served as a legitimate reprisal trope in dime novels, Hollywood Westerns, and computer games, rape went beyond the boundaries of national envisioning. *Custer's Revenge* demonstrated that violence had its own strict parameters of public acceptability. Massacres of "bad guys" and "angry Indians" were welcome, but violence in the form of rape was not in the traditional Western canon and thus off-limits to game players.

Custer, as war hero and national sacrifice, could also not be seen with his boots on but pants down. Popular culture mostly enshrined Custer as a war veteran and a victim of Indian aggression. In the decades immediately following his death, Custer emerged as national hero in the popular imagination. Captain Charles King, writing for *Harper's* magazine in 1890, described "a daring, dashing, impetuous trooper," with "unquestioned bravery," his one flaw being that the Colonel "lacked judgment."[25] A 1884 Beadle Boys dime novel cast Custer as "the flower of the American army, the brave, the gentle, the heroic, the people's idol."[26] A brochure for an Adam Forepaugh's Wild West Show in 1887 highlighted the show's detailed recreation of Custer's Last Stand, which depicted the "great moral drama" of the West and provided an emotive "chronicle of one of the saddest events in the history of Indian barbarities."[27] Forepaugh's show played at Madison Square Garden to much acclaim. The *New York Journal* enthused, "General Custer is being killed nightly while thousands of people witness the awful deed . . . The Battle of the Little Bighorn is a magnificent piece of realism, and stirs the beholder to the very soul." Moreover, on several occasions, dime novels teamed Custer with Buffalo Bill as valiant and determined justice seekers working together—a nineteenth-century version of Batman and Robin.

Custer's exploits made for easy translation to the twentieth-century silver screen. Filmed on the actual battle site, William Selig's *Custer's Last Stand* (1909) offered an allegedly "perfect reproduction" of events, while Thomas Ince's version, *Custer's Last Fight* (1912), employed local Indians as actors. Press material for the movie *Custer's Last Stand* (1936), directed by Elmer Clifton, emphasized the picture's realism and educational potential, a film that invited the audience to properly "relive the most dramatic days in all American history." The publicity projected Custer as "one of America's greatest heroes," and the film as one that "every teacher of history will want to see." Custer was imagined as the unfortunate hero, cast as part of a bigger story of inverted conquest in the West, where Indians threatened innocent white settlers.[28]

By the early 1980s, however, Custer's reputation had deteriorated as a result of new historical views that included Native interpretation. Custer instead symbolized the excesses of Western stupidity and conflict. In that light, *Custer's Revenge* appeared an awkward jump back in time to an era marked by ghastly violence and prejudice, succinctly summed up in General Philip Sheridan's phrase: "The only good Indian is a dead Indian."[29] As if the years had been rewound back a century, Custer was again portrayed as the hero, with the Native American as his adversary.

Custer's Revenge highlighted an international video games industry happy to depict American history as violent entertainment. Issues of morality or taste only arose with the most extreme of titles. Relatively few games explored notions of personal restraint or challenged players to differentiate between "good" and "bad" violence. Even fewer questioned conventional views of the West. While tracking down wanted gangs, such as the Daltons and the Petticoat Floozies, *Cheyenne* players had a duty to protect innocent bystanders from stray bullets. At the Last Dance Saloon, responsibilities included defending drinkers from falling masonry and from a wild, bar-busting bison that, once shot, transformed into a convenient bar snack. Meanwhile, Sega's *Bank Panic* (1984) invited the player to assume the duties of a sheriff named Hero, guarding a capacious bank from a never-ending queue of robbers. A dozen entry doors opened in quick succession to reveal earnest customers (male and female) carrying dollar-filled bags and gun-slinging bandits eager to make cash withdrawals. Although the Dixie soundtrack and frontier costumes evoked an earlier historical period, *Bank Panic* resembled a modern-day target range, with its two-dimensional pop-up figures. True to Western dictum, only by the hero outgunning his foe could law and order prevail over anarchy.

Video games represented the latest in a series of entertainment forms to tap the mythic Wild West, yet few noticed the abiding links between Western imagery, computer software titles, and the promotion of gun culture. While arcade machines never relied completely on Wild West backdrops, the Cowboys-and-Indians narrative contributed to a genre of interactivity predisposed toward shooting and killing. The imaginary Old West thus remained significant not only for its questionable presentation of nineteenth-century society, but also for its role in encouraging violent gaming. Western titles helped establish the gameplay mechanic of the "duel" and transformed the physical spaces of arcades into places of violent (albeit onscreen) confrontation. Immersing players in a digitally rendered Western story, *Gun Fight*, along with other early video games, overcame the limitations of its black-and-white game world by blurring the distinction between fiction and reality. Invited to take part in the violence on screen, players willingly participated in a form of digital gun culture: holding replica guns, taking physical poses, and firing as if faced by real enemies. Through their interactivity, video game shooters took the myth of a violent West one stage further than the dime novel and the Hollywood Western. Western shooters made onscreen killing in arcades an acceptable activity and helped tie Gamic America to repetitive acts of violence.

By the early 1990s, the two-dimensional Western shooter had mutated into three-dimensional titles based around massacring Nazis, monsters, and zombies. With id Software's *Wolfenstein 3D* (1992) and *Doom* (1993), followed by *Quake* (1996), *Unreal* (1998), *Half Life* (1998), *Medal of Honor* (1999), and *Halo* (2001), first-person shooters emerged as the dominant genre of video gaming.

In April 1999, Eric Harris and Dylan Klebold, both obsessive players of *Doom*, killed fifteen people at Columbine High School, in Littleton, Colorado. In response, US Senator Joseph Lieberman charged that computer entertainment glamorized killing. Video games arguably abetted a growing culture of violence in society. Some fifteen years after the first Western shooters, *Mortal Kombat* (1992), with its ultra-violent "Fatalities" (or finishing blows), and the killing marathon of first-person shooter *Doom* were the subjects of public outrage.

The Advancing Graphical Journey

The success of early video games relied on a combination of innovative programming and the presentation of simple ideas and visuals. Along with game play based around violent confrontation, programmers identified and transferred a range of potent symbols of the American West—rugged cowboys, desert vistas, and boozy saloons—to arcade screens. They created a recognizably Western landscape.

On aesthetic levels, the Old West was ideal terrain for arcade machine programmers. As Michael Johnson noted, the mythic, popular West was forged not by historians, but by "artists in visual media: painters, sculptors, filmmakers, and the like."[30] Technicolor images of "the West" were already firmly ensconced in mass culture before the advent of computer technology. The widely recognized Western iconography was well suited to 1970s and 1980s machines with limited graphical capabilities. A simple, pixilated version of the familiar cowboy was readily identifiable. Likewise, few could mistake a cactus, which was often all that was needed to signify a Western setting. With a lone pixelated cactus signifying nature west of the Mississippi, *Gun Fight*'s terminally empty landscape exploited a common misconception of the American West as stark and desolate. Moreover, the game's monochrome screen suited the black-and-white morality of Hollywood Western conflict, seen in movies such as *Shane*.

New arcade technology in the early 1980s expanded the gaze of the Western shooter. Color graphics brought an aesthetic richness to games. In contrast to the minimalist and monochrome landscapes of *Gun Fight* and *Boot Hill,* Sigma's *Wanted* boasted tiered panoramic views of green bushes and distant, sand-colored mountains. Bursting with vibrant colors in an almost abstract composition, Sigma's Western portrait loosely resembled Georgia O'Keeffe's emotive paintings of New Mexico or the literary landscape of Mary Hunter Austin's *The Land of Little Rain* (1903). The West appeared alive, natural, and fertile, rather than sullied, inconsequential, and dead.

Color video games in the 1980s depicted a region in transition, one that featured both prelapsarian beauty and signs of human industry. Reprising its role as a story backdrop, Monument Valley dutifully served as a staple background for successive arcade titles. An epic sense of scale and monumentalism marked titles such as *Wanted*. On-screen visual drama resembled Alfred R. Waud's explosive engraving of *Custer's Last Fight* (1876) and Frederic Remington's defiant, survivalist painting of *The Last Stand* (1890). Players looked out onto scenes of great spectacle. As if translating diary entries by the explorers Meriwether Lewis and William Clark on the inestimable numbers of deer, wolves, and antelope they encountered, the gamic West of the 1980s featured fauna in abundance.[31] Cattle thundered across the screen, vultures hovered in the air, wild mustangs brayed, and bleached bison bones littered the prairie. A signature of painter Charles Russell's romantic, eulogistic Montanan canvases, the buffalo skull attested to the harsh, testing qualities of the Western environment and the sorrowful, untimely passing of the frontier. Computer graphics likewise captured a bygone era of Euro-American discovery, reinventing a vanished West on screen, as well as tapping into profligate cultural interest in wilderness.

The 1980s digital West was also a visually dynamic landscape. Video games projected the Western terrain as a kaleidoscope of fast-moving frontier snapshots. In brief arcade bursts, gamers experienced an unbridled medley of animated boomtowns, pioneer industries, and Old West paraphernalia, amounting to a spectacular visual feast. The lightgun game *Cheyenne* offered encounters with classic locales through the scope of a large black rifle. By firing at representative pictures hung in a gallery, players chose to visit the Last Dance Saloon, explore a dingy gold mine, or hitch a ride on a stagecoach. A stirring cavalry bugle charge marked the onset of each stage. The Last Dance Saloon featured bartenders filling glasses, animal trophies hanging on the wall, and piano music playing in the background. The claustrophobic mine included rolling stock, shadowy villains, and dimming lights that led to a blackout only broken by gunfire. The stagecoach ride appropriately sped by with all manner of projectiles whizzing through the air (including bird droppings), while Apache Braves shouted "paleface" when dispatched with pixel lead. With its lively sounds, detailed settings, and fermentative atmosphere, *Cheyenne* perfected the dramaturgy of the Old West.

The speed at which players tackled (and conquered) the 1980s digital West put even the fast-moving emigrants of the California Gold Rush to shame. One image was replaced instantly with another, as if moving through a series of camera stills. Along with the train-based titles *Wild Western* (1982) and *Express Raider* (1986), Konami's *Iron Horse* (1986) portrayed a breakneck speed of progress. Reminiscent of John Ford's seminal movie *The Iron Horse* (1924), Konami's title highlighted the role of the railroad in Western development. With desert scenery racing past in the background, players moved through successive train carriages while engaging bandits in combat, hoping to thwart a hold-up of the US Pacific Railway in an allegorical battle for unfettered Western progress. The sound of steel wheels clanking on rail tracks promoted the sensation of forward motion. Whistling tunes conjured images of steam rising up from vintage locomotives and of melodic showdowns in cinematic Westerns.

A new, electronic frontier thus drew on nostalgia for classic images of the Wild West. It allowed gamers to look out across vast expanses, fight Indians, and charter steam locomotives. In arcades across the globe, gamers took part in a simplified version of the nineteenth-century Western story, made available to all nations, races, and classes. Regardless of the players' background or circumstance, all could take part in the journey. Marguerite Shaffer noted how wilderness travelogues in the 1910s and 1920s "encouraged tourists to reenact the nation's frontier past by camping and hiking."[32] Video games of the 1970s and 1980s furnished similar tourist experiences, of seeing and taming a new land.

Changing Wests

In the 1990s, the lure of the gamic West somewhat diminished. Identified with an older period of arcade games (and postwar American culture), the Western shooter seemed old fashioned and outdated. The simplicity of digital cowboy duels limited the games' appeal. Other more popular themes and genres gained popularity. Audiences welcomed the latest platform games on SNES and Genesis home consoles. Nintendo's *Super Mario* and Sega's *Sonic the Hedgehog* series, both family-friendly franchises with colorful, Disney-like cartoon characters, dominated sales. The Japanese survival-horror series *Resident Evil* (1996+) and the British adventure series *Tomb Raider* (1996+) appealed to mature audiences, especially those intrigued by transnational narratives and open to more than just an American authorial voice over game content. At the arcades, themselves in decline, *Virtua Cop* (1994), *Time Crisis* (1995), and *The House of the Dead* (1996), all set in contemporary times, offered fresh opportunities for shooting. The Gamer Nation seemed more in love with an Italian plumber addicted to mushrooms, lethal warriors engaged in *Mortal Kombat*, and a blue hedgehog with sports sneakers than with the classic American cowboy.

Programmers in the 1990s still applied the conventions and structures of the Wild West, but to new worlds and spaces. Programmers set up ambushes, duels, and showdowns, with lone and silent heroes meting out vigilante justice, but in contemporary or science-fiction settings. In the same way that the movie *Star Wars* (1977), with its frontier locales and gunslinger heroes, served as a space Western, a range of new games featured Western staples in disguise. The hugely successful stealth game *Metal Gear Solid* (1998), set in the 1990s, featured a mysterious hero (Solid Snake) sent into dangerous situations and facing a variety of foes in one-on-one showdowns, including a clash with Russian-born gunslinger Revolver Ocelot, who sported a duster, spurs, and Colt single-action firearms. Fighting, shooting, and exploration games all drew on familiar notions of honor, lonesome heroes, and the law of the gun.

The demise of the arcade Western seemed to replicate the vanishing of the Hollywood Western in the 1970s, albeit on a smaller scale. Tired and over-familiar, the Hollywood Western with its stories of manifest destiny and American glory felt at odds with a period marked by disillusionment over Vietnam, Watergate, and civil rights struggles. Video games also needed to reflect the contemporary mood. Unaccustomed to a diet of 1950s B-Westerns, audiences in the 1990s looked for different stories in their games than did players in the 1970s. Modern television programs depicted the 1990s West through images of crime, entertainment, and city life—themes picked up in the *Grand Theft Auto* video game series and other titles. A new Western iconography, marked by automobiles, shopping malls, and wilderness areas, shaped content. Titles reflected a growing cultural perception of the region as a realm of recreation and pleasure, or, to borrow Shaffer's phrase, "a landscape of leisure."[33]

New games based around the automobile employed the monumental scenery of the West as visual spectacle. Driving games resembled classic American road movies such as *Easy Rider* (1969) and *Vanishing Point* (1971). Players traveled along deserted highways with rugged peaks flashing by in their rear-view mirrors. The arcade title *18 Wheeler: American Pro Trucker* (2000) depicted a juggernaut ride from New York to San Francisco, in the same mode as the movie *Convoy* (1978). As the Asphalt Cowboy, players powered past swirling prairie twisters, dawdling Winnebagos, and Grand Canyon billboards. A paean to American freeway culture, *18 Wheeler* replicated the buzz of westerly travel along Route 66. The ability to complete the arcade game by recreating a 3,000-mile journey in less than thirty minutes defied logic as well as state speed limits.

Another product of the Japanese software giant Sega, *Crazy Taxi* (1999) explored the urban West by car. Arcade players assumed the role of a San Francisco cabdriver collecting fares from city-bred Westerners and foreign tourists, taking them to unique destinations, such as Candlestick Park and the Transamerica Pyramid, as well as to Kentucky Fried Chicken and Pizza Hut. By ferrying passengers to scattered locations, players steadily expanded their knowledge of virtual San Francisco, graduating from bewildered tourist to urbane docent. Tourists no doubt went through the same process in real taxi rides, just at a much slower speed. *Crazy Taxi* was a highly successful title both in the arcades and in homes.[34]

Harking back to shooters of the 1980s, computer hunting games highlighted the lingering appeal of marksmanship and hunting in the West. Titles such as *Deer Hunter* (1997) and *Cabela's Big Game Hunter* (1998) prided themselves on realistic portraits of American sport hunting. Boasting three-dimensional graphics, varied weather conditions, live-video footage, and "real animal sounds," *Cabela's Big Game Hunter III* (1999) offered opportunities to bag elk, bighorn sheep, wolves, and grizzlies. A fan of *Deer Hunter* established Antler Creek Lodge on the World Wide Web, an attractive Western ranch complete with trophy room, Indian totem pole, and a recreational vehicle. Visitors to the lodge gazed at "kill" pictures taken from *Deer Hunter*, noted valuable information on prey and guns, and entered online hunting trips. Other players found hunting in the virtual wilderness a decidedly drab affair. Disappointed by the lack of action in *Big Game Hunter*, a reviewer for VirtualOutdoorsman.com noted that it was "a good game until you actually start hunting." A guest reviewer for *Gamespot* declared *Deer Hunter* "so appallingly bad that I suspect that it was actually created by the Sierra Club in an effort to dissuade people from hunting."[35]

The few titles that continued to tackle the nineteenth-century American West faced the choice of whether to regurgitate simple cowboys-and-Indians stories of the past or offer something fundamentally deeper and more challenging. Giving the player directorial control over classic Wild West vignettes, Infogrames' *Desperados: Wanted Dead or Alive* (2001), *Desperados 2: Cooper's Revenge* (2006), and *Helldorado* (2009)—three real-time strategy games—marked a return to the old West, but with a new sophistication in the digital narrative based around the concept of "directing stories." *Desperados* engaged with classical storytelling by offering twenty-four mission-based "best western adventures" for gamers to navigate, but with the player cast as a director in each film playing out. The official press release by Infogrames exploited plenty of cowboy clichés, asking players to "dig out your poncho and dust down your Stetson," and "saddle up for the wildest adventure this side of the Alamo!"[36] However, the roster of six playable characters was remarkably revisionist, including an ex-slave, a Mexican, an Asian American, and an empowered female poker player by the name of Kate O'Hara. *Desperados* allowed players significant freedom to create their own historic West as postmodernist digital authors.

Programmers of *Gun* (2005), meanwhile, chose a more conventional route, updating an old action story with modern, three-dimensional graphics. Representative of the increasing crossover between film and video game, Activision's *Gun* boasted an array of Hollywood actors lending their voices to the title, including Kris Kristofferson and Dwight Schulz. Relatively popular, it sold 225,000 units. The clichéd story courted controversy with its representation of Apache Indians. The Association for American Indian Development petitioned for a public boycott of the title based on the game's "derogatory, harmful and inaccurate depictions of American Indians," citing misinformation about Native traditions, an encouragement to players to scalp Apache in order to progress, and, on a wider level, the promotion of Indian genocide within the game. Activision apologized for any offense caused. However, in a public statement, the corporation suggested that it was merely repeating what had gone before: "the game's depiction of historical events . . . have been conveyed not only through video games but through films, television programming, books and other media."[37]

In 2004, Rockstar released *Red Dead Revolver*, a third-person adventure set in the 1880s Wild West, inspired by Italian Westerns and the Capcom game *Gun Smoke*. The title was well received in the gaming press and sold almost a million copies. With a budget exceeding $80 million, Rockstar San Diego programmers worked for five years on a successor title. Released in 2010, *Red Dead Redemption* promised new depth to video game historical fiction. The title sold over fifteen million copies. *Red Dead Redemption* related the troubled story of ex-gunslinger and gang member John Marston as he sought justice and the return of his family. The game consisted of a series of action-based missions, but also offered an "open world" (or West) to freely explore. Activities included horse taming, gambling, wildlife hunting, and herb collecting. Drawn extensively from Western film, as well as archival research at the Library of Congress, the *Red Dead* West featured realistic settings, props, and weather conditions, including spectacular thunderstorms. Combining aural and visual effects, *Red Dead* offered an unbridled sense of immersion and realism. A reviewer for the *New York Times* enthused, "[John Marston] and his creators conjure such a convincing, cohesive and enthralling re-imagination of the real world that it sets a new standard for sophistication and ambition in electronic gaming."[38] The digital West had evolved from simple stick cowboys to virtual realism.

Red Dead Redemption also explored the notion of a complicated West. In contrast to older games that glorified the cowboy kingdom,

Rockstar's title presented a troubled, complicated, and dying region in its "twilight years." Rockstar promoted its title as loyal to history and the "True West."[39] Set in 1911 on the Mexican border, the game paralleled the closure of the frontier with the advent of new twentieth-century technology. The clash of old and new could be seen in the *Red Dead* landscape. Traditional frontier towns looked antiquated and barbaric when compared with the industrializing city of Blackwater on the Great Plains (based on Blackwater, Missouri), which featured gas lamps, cobbled streets, telephone poles, automobiles, and a modern police force. The intrusion of the automobile in particular provided a visual contradiction conspicuous to the player. Director Sam Peckinpah explored a similar paradox in the opening sequence of the film *Ride the High Country* (1962). In the sequence, a police sheriff shouts, "Watch out, old timer" to Steve Judd (played by Joel McCrea), who is riding his horse, oblivious to all around him, as a Ford Model T automobile enters the main street. In *Red Dead*, the character of Marston harnessed the technologies of both old and new West, adept with both the lasso and hunting knife, while employing the latest dual-prism binoculars to spot his enemies.

The game referenced a number of cult movies in its illusion to a darker and dying West. In the 1953 George Stevens movie *Shane*, Shane is a gunfighter without purpose, much like Marston, lost and alone in a world that has moved on from six shooters, who becomes a hero grudgingly brought back into service. Marked by nihilistic but beautifully shot violence, the world of *Red Dead* stylistically reprised Peckinpah's *The Wild Bunch* (1969), including an in-game Gatling gun sequence that seemed directly taken from the motion picture.

Red Dead portrayed a West on its death bed, full of deceit and damnation, overtaken by twisted crooks and brutish mechanisms. Depicting the fate of the West as symbolic of an equally corrupt and violent nation, Rockstar's title offered acerbic criticism of the trajectory of the United States. As a character in the game, oil salesman Nigel West Dickens, warned, "This is America, where a lying, cheating degenerate can prosper."

Reading the Digital West

From *Gun Fight* to *Red Dead Redemption*, the computer entertainment industry clearly developed a more sophisticated presentation of the historic West over four decades. Whereas early arcade machines sold uncomplicated cowboy chronicles, with the video game industry simply providing new technology to preserve an old chimera, by the 2010s, *Red Dead Redemption* offered a far more rich, filmic, and challenging West.

The success of titles across several decades of play reflects the longevity and endurance of Western mythology in US popular culture. In 1996, Michael Johnson commented on the supreme fashionability of the West; despite Hollywood producing fewer Westerns, the West continued to be "mythologized, fetishized, stereotyped, commodified" on television screens, on magazine covers, and in clothing stores.[40] Shackled to the same ideology of a "cowboy nation," the video game Western profited from the same, ever-popular gunslinger fantasy. Following dime novels, Wild West shows, and Hollywood movies, the video game Western marked the next step in the cultural processing of the American West. Video games joined what Richard Aquila once labeled the "pop culture West" of action stories.[41]

The West imagined in Gamic America offered a number of tweaks to the traditional narrative. Especially in its early incarnations, the arcade game supplied the most pure and reductive take on history for some decades. In a dialogue between the dictates of new technology and established cultural mores, programmers transformed the West into an easy-to-digest, five-minute story. The mythology of the frontier—as Richard Slotkin put it, "progress through violence"—became a series of stages of game play. Players moved successively forward by fighting Indians and claiming digital territory. National redemption and renewal of spirit came in the form of points on the screen. In the process, white agency triumphed, and a range of other actors disappeared from the historical stage. The linguistic and literal mythology of the West became marginalized, and the visual mythology of the region became stronger than ever. Given the success of the Hollywood Western and the dominance of the image in the late twentieth century, such a route seemed predictable. However, digital space comprehensively diluted the Western story.[42]

Games also strengthened the conception of the American West as a violent domain. Titles such as *Boot Hill*, *Blood Bros.*, and even *Desperados* and *Red Dead Redemption* presented the Western endeavor as a series of continuous and gloriously iconic gunfights—propagating the most important myth of American history as violence par excellence.

While Hollywood had popularized a gun-toting West on screen, the game industry took the next step by providing its audience with targets and lightguns. Fusing together material culture and digital technology, players gripped replica rifles and fired at pixel cowboys. Games asserted the popular myth of the West as violence and coupled that myth with new technology and a younger audience. While *The Oregon Trail* and *Civilization* presented America as a tactical, nation-making enterprise, full of calculation and resource management, Western video games offered an American experience that was all about visceral displays of brute force.

Early Western video games also revitalized the sense of the West as a story of victory. Gamic America sanctioned not just a visit to a past, but also a decisive win over it. Games created a past America for players to explore, then ultimately conquer, and a world where every gamer could be a cowboy and shoot an Indian. Games visualized American history as a game to win, a troubling concept given the massacres and wars that marked the real nineteenth-century experience. Like Disneyland's Frontierland, the Gamic West told an entertaining story to suit its white, middle-class guests, but one with only a loose hold on reality.

As a result of larger budgets, more ambitious stories, and more creative programming, a few games studios in the 2000s began to test the story of the West that the industry had formerly embraced and codified. Offering a more expansive and detailed simulation of events, *Desperados* and *Red Dead Redemption* crucially questioned the reliability of the narrator, and also the player, as sticky historical situations played out. The story of the West suddenly became problematic, postmodern, and open to interpretation. Rather than follow a set path through an action-based arcade game, players faced a choice of routes, with differing moral implications. The omnipotent power of the Westerly dream was thus, if not totally rejected, at the very least questioned and prodded, with American history left intriguingly open and malleable to new interpretation.

3

Cold War Gaming

CONCEIVED IN THE BASEMENT of the Electrical Engineering Department of the Massachusetts Institute of Technology (MIT) in 1961, *Spacewar!*, widely considered the first computer game, demonstrated the potential of supercomputers for play as well as calculations. With the graphical programs *Mouse in the Maze* and *Tic-Tac-Toe* already working on Digital Equipment Corporation's PDP computer, a group of MIT students, led by Steve Russell, set about crafting a fast-paced action game. They created a futuristic game in which players controlled two spaceships engaged in combat on a monochrome view screen. *Spacewar!* was a product of MIT research culture, student creativity, and excitement about the possibilities of new technology.[1]

Spacewar! was also a product of the Cold War. Its genesis came at a time of peak hostilities between two superpowers vying for global dominance through the space race. In August 1957, the Soviets tested the first intercontinental ballistic missile (ICBM), the R7, which travelled 3,700 miles before crashing into the Pacific Ocean. With a modified R7 delivery system, Sputnik, the world's first satellite, launched into space in October 1957. In April 1961, cosmonaut Yuri Gagarin became the first human in space, successfully orbiting Earth in his Vostok spacecraft. That same year, President John Fitzgerald Kennedy committed the United States to putting a man on the moon by the end of the decade.

Spacewar! captured the sense of technological conflict at the heart of the Cold War. Its ludic content reflected not just superpower rivalry, but also the quest for technological dominance marked by nuclear weaponry, space travel, and advanced missile technology. The game reflected new trends in technology, science, and culture. *Spacewar!* showed that video games, from their very inception, reflected, recoded, and commented on technological progress, societal hopes and fears, and real-life events. Rather than occupying their own creative vacuum, computer games engaged with the world outside them. Computer games not only re-imagined the American past, such as the conquest of the Wild West, but also actively engaged with the present. Moreover, they produced new ways of engaging with salient political issues.

Spacewar! transformed the space race into simple computer gameplay. It brought the ICBM rockets, spacecraft, and dueling powers of the Cold War together on a single view screen. The game represented the intense sparring of the United States and Soviet Union as two opposing ships facing off in the vacuum of space, with the sole imperative to destroy one another. A needle and a wedge flew across the screen, with stars in the background. Players could rotate the ships, accelerate their speed, jump to hyperspace, and fire ICBM missiles. In early scripts for the game, programmer J. Martin Graetz related, "the ships would have a supply of rocket fuel and some sort of a weapon: a ray or beam, possibly a missile."[2] The final choice of ICBMs situated the game in the contemporary atmosphere of the Cold War. Like their real-world counterparts, ICBMs shot across the sky and exploded on screen. To further realism, programmers added a "night sky" code, nicknamed "Expansive Planetarium," along with a realistic gravity program. In a tightly orchestrated computer duel, the MIT programmers had gamified the Cold War.

While the technology was new and experimental, ideas for *Spacewar!* emerged from an established Cold War gaze. The game world drew on older fictions. Graetz cited the gung-ho science fiction novels of Edward E. Smith as influential in the game's design. First published in 1946, Smith's *Skylark* series was popular with the MIT team. In *Skylark*, "Smith mixed elements of the spy thriller and the western story (our hero is the fastest gun in space, our villain the second fastest) with those of the traditional cosmic voyage."[3] The novels idealized a strong American character, an identity soon caught up in the foreign politicking and saber rattling of the Cold War, exemplified in the US cowboy versus the Russian bear. Smith developed plots around the atomic bomb to set them within a distinctly Cold War frame. Graetz also cited Toho Film Studios as inspiration for *Spacewar!*. Toho produced a range of monster and special effects movies, the most famous being the *Godzilla* series. Graetz and company enjoyed the "cinematic junkfood" of Toho productions, marked by their "oceans of rays, beams, and explosions."[4] However, some Toho movies articulated far more than a popcorn agenda. The original version of *Godzilla* (1954) subtly critiqued the US nuclear testing program of the period and highlighted the horror of the nuclear age. Its opening scene referenced the poisoning of the crew of the Japanese trawler *Lucky Dragon* during Castle Bravo tests in March 1954; Godzilla later ravaged cities on a scale equivalent to the damage done by atomic weapons to the Japanese mainland. Godzilla was widely taken as a metaphor for Japanese suffering inflicted by the bombing of Hiroshima and Nagasaki. A carefully edited, American-friendly version of the original movie appeared in US markets in 1956.

The making of the game also connected to the Cold War in tangible ways. While working on *Spacewar!*, programmer Wayne Witanen, a reservist in the US Army, left for active duty connected with the Berlin Wall blockade of October 1961. The computer hardware itself had practical ties with the military machinery of the Cold War. The PDP-1 by Digital Equipment Corporation, designed at MIT's Lincoln laboratory and introduced in 1959, sported solid-state technology, 4k of memory, and a $120,000 price tag. Measuring 8-feet-deep by six-feet-high, it was considered small by the standards of the time. The North American nuclear and defense industry speedily co-opted the PDP-1, with Lawrence Livermore Laboratories and Atomic Energy of Canada placing early orders. While the PDP at MIT simulated ship-to-ship combat in *Spacewar!*, PDP national defense computers calculated the logistics of two superpowers locked in conflict.[5]

Whether through military-sponsored calculations of battlefield exchanges or games programmed in MIT basements, computer programs were fundamentally connected with the Cold War experience. While rocket-powered missiles were the most visible, and thus most feared, aspect of Cold War technological warfare, computers operated behind the scene, controlling military operations from afar. Historian Stuart Leslie pronounced, "For better and for worse, the Cold War redefined

American science."[6] Conflict shaped technological advances, the imaginary front line demanding (as well as funding) major industrial innovation. Part of the military–industrial complex of the Cold War, computer technology served a burgeoning defense economy. Computers operated and often automated many of the core defense systems of the next half-century. Early UNIVAC computers ran the Command Center Processing System of NORAD at Cheyenne Mountain Nuclear Bunker, in Colorado, which was responsible for the nation's atomic arsenal. Computers similarly ran sophisticated political–military simulations at the Pentagon.

Filling entire rooms, early computers resembled mechanical shrines to data crunching and mathematical rationalism. Computer simulations and programmable scenarios proved crucial to Cold War planning. Computer war games shaped conceptions of the Cold War, with the conflict seen through the mediated form of machine technology. While in the case of *Spacewar!*, the PDP facilitated pretend conflict between two players in their virtual spaceships, military computers had the potential to orchestrate real-life conflict between the superpowers. They could make war happen. Placed inside missile silos and communication devices, computers allowed war games and simulations to progress to real battlefields, turning the Cold War to "hot." Both nations developed their own "doomsday machines" capable of instigating Armageddon without presidential authority or human oversight. The Soviet's Dead Hand (or Perimeter), operational in the 1980s, was capable of independently activating a nationwide counterstrike based on seismic activity and radioactive data.[7]

Along with film, music, and literature, video games from the 1960s through the 1980s explored contemporary Cold War themes. A small but important number of titles consciously tackled the clash between superpowers, articulated nuclear fears, and even promoted protest agendas. Arcade machines such as Atari's *Missile Command* (1980) and Irem's *Red Alert* (1981) gamified conflict. Some titles reproduced narratives established in conventional media; others pioneered new ways of seeing and interacting with the Soviet threat. The digital visualizing of the Cold War often coupled popular angst with an action-based, military-like outlook. Cold War games designed for the home user borrowed from what people imagined DEFCON control rooms and doomsday machines looked like. Game screens mimicked the appearance and feel of military hardware in nuclear bunkers. The war games being played at home resembled the war games being played at US military bases, player and soldier alike tackling a range of doomsday scenarios.[8]

Missile Command and the Cold War Nightmare

In the early 1960s few Americans had seen a computer game, yet alone played one. By the early 1980s, however, America had become a fully-fledged gamer nation. No longer confined to MIT basements, video games could be found in shopping malls and living rooms. *Pac-Man*, *Space Invaders*, and *Asteroids* were widely recognized titles. Now capable of reaching vast audiences, video game content had the potential to shape minds as well as hone play skills.

At the same time, the Cold War was reaching its peak. The Soviet invasion of Afghanistan in 1979, the US boycott of the Moscow 1980 Summer Olympics, and a shift in Western foreign policy following the election of Ronald Reagan in the United States and Margaret Thatcher in the United Kingdom heralded the return to a more heated phase of East–West relations. A period of détente gave way to a new aggressiveness in superpower rivalry. Heightened tensions between the United States and the Soviet Union brought about a fresh escalation in the nuclear arms race. Nuclear war seemed perilously close.

In June 1980, Atari released the arcade machine *Missile Command*. The game situated the player as a missile controller responsible for the defense of multiple cities undergoing a nuclear attack. Atari promoted *Missile Command* to trade clientele that year as "the ultimate action experience, the ultimate profit weapon."[9] The title was immensely popular. Around 20,000 units shipped. A home console conversion for the Atari VCS became the best-selling video game the following year. The company went on to sell 2.76 million game cartridges.

A product of renewed Cold War anxieties, Atari's title explored the worst-case scenario of missile exchange: an all-out assault by the Soviets. Contemporary fears of conflict directly inspired the game. As lead programmer Dave Theurer noted, "*Missile Command* embodied the Cold War nightmare the world lived in."[10] For several decades, the US military had explored the notion of protecting key American targets from Russian attack. In 1963, Defense Secretary Robert McNamara backed Sentinel, a missile defense system for the continental United States, but issues of practicality as well as its potential to undermine the

Strategic Arms Limitation Talks between the superpowers, crippled the project. In the 1970s, the US military developed a program whereby Spartan interceptor missiles would protect strategic missile silos, ensuring counterstrike capability. A range of flaws alongside treaty restrictions led to its sole deployment at Stanley R. Mickelsen Safeguard Complex in North Dakota. Four years later, Lieutenant General Daniel O. Graham, backed by President Reagan, conceived a successor, code-named High Frontier. Based on ICBM missiles and space lasers eliminating all incoming Soviet missiles, High Frontier promised a radical solution to the sizable challenge of protecting the nation from attack. Renamed the Strategic Defense Initiative (SDI) in 1983, the science-fiction imagery of lasers on a "high frontier" lent the scheme the popular moniker of Star Wars.[11] SDI could equally have been dubbed *Missile Command*. The US defense system mirrored the game in its reliance on high technology and a battle taking place in the sky. It engaged with the same threat of Soviet missiles bound for US land, and it offered a similar solution based on missile counter-deployment. *Missile Command* thus captured contemporary military and scientific discourse on the defense of home and country. Atari's title proved a timely take on a key debate of the 1980s: how to survive nuclear war.

A magazine article about satellites and radars provided the brief for the Atari game. President of Atari Coin-Op Gene Lipkin passed the article onto team leader Steve Calfee. Calfee gave the task to programmer Dave Theurer: "Make me a game that looks like this."[12] Early design briefs focused on radar screens, missile trails, and interceptors. Theurer recalled that the aim was to make a game "where the Russians are invading the USA, and there's this radar screen and can see the missiles coming in and you have to defend the country."[13] At the start, programmers hoped for a relatively realistic depiction of nuclear conflict. The design team consciously situated *Missile Command* as a simulation of a nuclear attack on the player's homeland, with all the pressures and anxieties involved. The game was to be grounded in real-world geography, with the player motivated by the defense of his or her home territory. In the initial design, the game was to be set in California, with the six coastal cities of Eureka, San Francisco, San Luis Obispo, Santa Barbara, Los Angeles, and San Diego under attack.[14] The Atari team considered the potential for arcade machines to be region-specific, with the California locations replaced by local landmarks.

Working on the game generated a series of Cold War fears among the team. Theurer found it "pretty scary" and suffered reoccurring bad dreams. Calfee noted, "Everybody I knew who got into the game had nightmares about nuclear war."[15]

Due to the technical limits of hardware and wariness over public receptivity, the game's content shifted over the course of development. Designers eschewed social realism for symbolic representation. Concern over a crowded screen and over-complicated gameplay led to the axing of radar, supply railroads, and factories. Technological limits further restricted what could be put on screen. Cities and missiles were cast in rudimentary pixelated graphics. Fear over the impact of game content on public attitudes also led to revisions. Atari dropped the working title "Armageddon," and California-specific cities disappeared in the shipped arcade machine. In the 1981 Atari VCS console version, all references to America under threat were removed. The console manual instead provided a science fiction backstory, with America renamed Zardon and the Soviets represented as the alien species Krytolians. Only subtle analogies remained, with descriptions of Zardon as a place "void of crime and violence . . . happy and harmonious," matching an idealized version of the United States. Cruise missiles and antiballistic missiles survived the adaptation; Cold War technology somehow transcended space and time.[16]

In its final form, *Missile Command* offered a focused and elemental vision of nuclear attack. The title provided an unfussy symbolic presentation of Cold War conflict. With little contextual information or explanation, players navigated an atomic strike, unaware of the political backdrop, or even who or what launched the missiles. Rudimentary graphics and a simplified narrative reduced the Cold War to bare essentials. Programmers cast the Cold War as a simple action plan: defending cities from imminent destruction. Atari's work synthesized and deconstructed classic Cold War fears and concerns. Multiple missile exchanges and high scores complemented the real-world focus on which superpower had the most weaponry. In particular, the focus on the city, the missile, and the mushroom cloud captured the core iconography of nuclear war.

For decades, Cold War culture highlighted the vulnerability of American cities and the need for their defense. In the 1950s, the Federal Civil Defense Administration (FCDA) promoted the need for citizens to take up civil defense, build shelters in their homes, and gather supplies to survive possible attack. The FCDA even simulated Soviet invasions in Operation Alert, a drill held routinely across US cities. In 1953 and 1955, two small fake towns (nicknamed Doom Towns) were exploded at Nevada Test Site as simulations of nuclear attack. In 1959, a couple even honeymooned in a bomb shelter. The need to defend the American homeland was a longstanding concern that predated Reagan's Star Wars by some margin. The imagining and simulating of atomic onslaught thus existed in both official circles and in the public mindset long before *Missile Command*.[17] Atari transformed that imagining into a game. It also changed the playing field. Whereas past simulations and drills allowed participants only the scope to prepare for inevitable destruction, *Missile Command* offered the player the option to stop missiles from reaching cities in the first place. The player could take a proactive role in determining the fate of the nation. Rather than helplessly preparing for the end, players could sacrifice or save American cities with a simple press of a button.

Since the introduction of ICBMs in 1957, missile technology had shaped the fabric of Cold War conflict. As the major delivery system of the atomic bomb, the ICBM determined strike ranges and affected geopolitics through the need for proximate launch sites. In the early 1980s, Reagan's escalation of the US nuclear arsenal, to be delivered by American cruise missiles, dominated newspaper headlines. Along with skybound Soviet R-series projectiles, cruise missiles launching from silos became the dominant visual marker of the looming danger of nuclear war. Atari's title captured the frightening scenario of missiles homing in on the American heartland. It showed a spread of nuclear weapons dropping from the sky. The sheer number of missiles onscreen fed a broader sense of a nuclear arms race out of control, of spiraling numbers and the likelihood of massive overkill. However, in *Missile Command* the same technology that threatened death and destruction was also the ultimate cure—missile versus missile. This same logic permeated Cold War discourse: meeting threat with threat, technology with technology.[18]

On reaching their targets, the game's pixel missiles exploded, destroying cities in an instant. However, if intercepted early, the missiles simply detonated on route, marked by a huge dispersal cloud that lit up the sky and engulfed other missiles, rendering them disabled. Atari programmers cleverly employed a classic gameplay form (the chain reaction), placing it within an existing scientific thread of atomic chain reactions. Chain reactions advantaged the player. One well-targeted American missile could destroy multiple Soviet ones thanks to this process. Successful games were marked by the dominance of atomic clouds onscreen. Multicolored and morphing in size, the clouds made the game an ever-changing and exciting visual spectacle. The pixelated animations provided a continual flashing repeat of the core image of the Cold War: the nuclear cloud.

Ever since the Trinity test in 1945 at Alamogordo, New Mexico, the mushroom cloud dominated the atomic gaze. As Scott Kirsch notes, "a picture of the familiar mushroom cloud" became "the logo for the atomic age," with the Atomic Energy Commission in the 1950s cleverly disassociating the "sublime, visual qualities" of the explosions from the "singular devastation of landscape" beneath them.[19] *Missile Command* tapped this established popular fixation with the mushroom cloud. A fundamentally frightening image of the Cold War served as an icon universally recognizable across arcades. Players were caught in an act of visual repetition and of continual cloud making.

Despite the revisions to the initial brief, Atari's title retained much of its Cold War imagery on delivery to arcades in the early 1980s. The final product visualized a Cold War world of missiles and machines, quick reactions and deadly threats. Cabinet art featured the red button, a destroyed city, a mushroom crowd, incoming atomic missiles, and US military personnel. With its use of ICBMs and Red Alerts, and its frequent reference to annihilation, the game incorporated Cold War terminology into the pixelated landscape. While the VCS console version used an alien attack scenario, an Atari 400/800 home computer iteration, like the arcade machine, pushed real-world associations. Assuming that computer owners were mostly adults, the 400/800 version offered a mature and contemporary narrative. Box art featured the familiar triangular logo for Civil Defense, while a game manual offered a detailed Cold War backstory.[20]

At home or in the arcade, gameplay effectively mapped the dramaturgy of Cold War conflict. *Missile Command* simulated one potential attack scenario after another, and forced the player to react. The game functioned as a pixelated fantasy of Armageddon. Atari encouraged a process of blurring the boundaries between the digital realm and real life. One Atari manual introduced the story of *Missile Command* as a military buildup and posed to the player, "Could this build up really be exercises? War games? Don't kid yourself, this is the real thing. It's only a question of time."[21] Actual missile commanders working at stations across the country seemed not that far removed Atari players. Both gamers and military personnel faced the same dangers and precarious situations: "You press a button on the console in front of you . . . enemy on the monitor screen. 'Fire.'" Standard arcade game dynamics—fervent button pressing, fast thinking, quick bouts of concentration, action-based screens—were all used to create an immersive drama. The frenetic quality of *Missile Command* induced a sense of adrenaline and anxiety in the player. The need to continually dispatch missiles to block a never-ceasing bombardment of lethal nuclear weapons created a sense of pressure and focus. The sense of panic, the imminent danger, and the fear of cities about to be lost provided a gamic take on real nuclear fear. Like the real-world siren warnings designed in the 1980s to announce that the Cold War had gone hot, each game screen began with an urgent call to defend cities in the face of attack. Sharp reaction times mattered in Atari's title, as they did in military simulations of Cold War scenarios—only by instant response might Armageddon be prevented. An itchy trigger finger and a clear mind in the face of disaster seemed crucial to both real and imagined conflicts. Thus, *Missile Command* articulated broader fears of destruction, annihilation, and survival in the nuclear age.[22]

The game also granted players a role in Cold War events where normally none existed. For most Americans, the Cold War was something distant, inaccessible, and out of their hands. It was a distant war game played by global superpowers. Presidents made the decisions, and technology delivered mass death. These broader cultural understandings influenced the design of *Missile Command*. Atari programmers toyed with the status quo. A cathartic quality came from living the fantasy of not only playing a commanding role in a Cold War gone hot, but also saving American cities and American lives. The player emerged as a Cold War hero, a valiant soldier doing his or her best to defend home and country.[23]

The gamic Cold War hero nonetheless had limits. The need to sacrifice cities and ultimately, at the end of the game, face defeat, always undermined the level of heroics. A sense of desperation, even futility, lurked within the game world. Defending all five cities was impossible, placing the gamer in the difficult position of sacrificing some to save others. While players could continue defending cities for hours, eventually the situation became untenable. The conflict was unwinnable. Instructions for the arcade machine stated simply, "The game is completed when all the player's cities have been totally annihilated."[24] *Missile Command* provided no end scenario other than defeat and no game screen other than a statement of numbers saved and lost. That Atari's title was unwinnable placed it alongside many arcade titles of the period: success in most early video games was based on high scores rather than completion. Games ended by the player succumbing to the ghosts in *Pac-Man* or the alien swarms in *Space Invaders*. Computer games focused on challenge, repetition, and score charts, not on completion or victory.[25]

The unwinnable nature of *Missile Command* suited the protest rhetoric of the Cold War. Nobody could win in a global nuclear war. Gameplay of the period thus reinforced social wariness toward conflict. The game highlighted the impossible challenge of facing atomic bombardment and offered a poignant message: nuclear conflict, once started, would decimate America. Designers coded political commentary within the game; people learned through experience and through play, that, whatever strategy they employed, nuclear attack could not be countered. As Theurer noted, "[the] whole point of the game was to show that if there was ever a nuclear war, that you would never win. So that's why there is no end."[26]

In another political statement, the team refused to show missiles attacking the USSR, basing their stance—and morality—on defensive posturing. Players became actively involved in saving cities, not destroying them; they could stop missiles, but not send them. The "enemy" remained clouded and distant. Operating outside both traditional game logic (of vanquishing the enemy in a duel, as in Atari's classic *Combat*) and the framework of Cold War logic (of mutually assured destruction), *Missile Command* challenged and subverted conventions. It presented a pacifist line. Clearly, early video games offered more than just play.[27]

Writing for *Film Comment* in early 1983, Mike Moore situated *Missile Command* as one of a range of popular games that attracted New Yorkers to a local arcade in Times Square, a dark space that the journalist dubbed "as seedy as seedy gets."[28] For Moore, the arcade games had a common theme: calling on the player to meet the "appearance of some new enemy." Moore hypothesized, "If defending the nation were possible by flipping a lever, spinning a sphere and pushing buttons, America has an army in training." For some players, titles such as *Missile Command* likely took on the feeling of fighting for America. For others, the ultimately unwinnable nature of the conflict raised broader questions about real-life war. For those absorbed in the onscreen action and constant button pressing, both the Cold War scenario of *Missile Command* and Atari's subversive message may have been missed. The activity of arcade gaming, marked by quick reactions and high scores, may have undermined the relay of any contextual information. The game remained just a game. Symbolic presentation and the lack of narrative limited Cold War association, with experiences of player anxiety some distance from any sensation of nuclear fear. While programmers had nuclear nightmares because they intimately understood the coded meaning of the game, more casual gamers may have simply suffered from an itchy trigger finger. The greater question was, could complex emotional reactions, such as nuclear fear, be provoked by 8-bit programs and stick figures?

Missile Command helped set the tone of Cold War computer gaming. It portrayed a global nuclear theater on the brink of a lethal missile exchange. It cast doubts over Cold War political decision making. In stark contrast to writer and director John Milius's celebratory movie *Red Dawn* (1984) starring Patrick Swayze, *Missile Command* offered no sense of gung-ho patriotism.[29] Rather than provide a ludonarrative of America triumphing over the Soviet Union, Atari programmers depicted the nation as inevitably losing to enemy fire. Atari's coding of the Cold War as a series of simple icons depicting missiles, bases, and cities endured in other titles. The game also situated the player as an overseer—a button presser, a reactor—rather than an on-the-ground participant (or victim) of radioactive contamination. Thus, the title provided an important framing for interaction. The gamic Cold War was impersonal, distant but deadly. Ultimately, this served to limit the player's experience. Depictions of nuclear conflict made the Cold War something that the player observed but did not fully interact with. The full repercussions of a radioactive war never reached the gamer. In that sense, games perpetuated some of the ignorance and denial of the period, aspects that dramatic movies such as Lynne Littman's *Testament* (1983) desperately tried to counter by involving the viewer in on-screen family trauma and community death.

A similar style of symbolic representation of the Cold War featured in other early 1980s video games. The arcade machine *Red Alert* (1981) by Japanese company Irem transformed the Cold War into a *Galaxians*-style shooter, with the player downing enemy warplanes in order to prevent nuclear attack on various Western capitalist countries. One of the first games to feature digitized speech, *Red Alert* included the chilling announcement "Red Alert. Enemy Aircraft Approaching. United States." If the player succeeded in defending a country, "Peace Forever" lit up the screen.

B-1 Nuclear Bomber (1980) for the Apple II computer turned the Cold War into a text-based flight simulation. The player flew the bomber via a series of simple binary commands, such as setting a course (CA) or altitude (AL), taking evasive action (EV), or bombing (BO). The aim was to reach one of a series of Soviet destinations and successfully deliver a nuclear payload. With just the background noise of air flight and a computer asking questions, the game provided a limited sensory experience of conflict. With just the occasional text on a dark view screen, the player could nonetheless imagine flying a mission in a Cold War world of radio silence and secrecy, lone survival and imminent danger. Technological restraints thus helped create a powerful visceral experience. The onscreen announcement "Hot War. Hot War. Hot War" shocked the player precisely because of the scarcity of information.[30]

Playing *WarGames*

The Cold War of the 1980s provoked all kinds of cultural comment, satire, and protest. From the television movie *The Day After* (1983), which depicted an attack on Lawrence, Kansas; to the popular song "99 Red Balloons" (1983) by Nena, protesting NATO missile deployment in Europe; to Alan Moore's dystopian comic *Watchmen* (1987), with its retired American supermen and an impending World War III, various media articulated public concern. Computer games likewise connected with the Cold War, on some levels more than anticipated. In 1983, MGM released the movie *WarGames* at the box office. An early cyber-thriller directed by John Badham, *WarGames* depicted a teenage computer hacker, David Lightman (Matthew Broderick), playing what he thinks is a new game, but is actually a war simulation with the US Air Force's WOPR (War Operation Planned Response) computer, nicknamed Joshua. By accepting the challenge to play a game, Lightman unknowingly instigates a countdown to nuclear war. Accused of spying, Lightman attempts to convince authorities otherwise, while trying to get WOPR to cancel Armageddon. The finale involves the NORAD control room, DEFCON countdowns, and a computer close to activating nuclear launch codes and inviting doomsday.[31]

WarGames brought together two cultural anxieties of the early 1980s: computer hacking and nuclear war. It forged a worst-case scenario by fusing Cold War fear with anxiety over a new computer age. As Vincent Canby for the *New York Times* noted, the film "immediately connects with our nightmares about thermonuclear war and a world ordered by unreliable computers, it grabs us where we're most vulnerable."[32] *WarGames* was a huge success; it netted $79 million and received three Academy Award nominations.

WarGames's popularity also highlighted the growing significance of video games and computers in American society. That computer games could now inspire narrative in mainstream Hollywood movies indicated a new interest in and acceptance of them. The Disney movie *Tron* (1982) tackled similar themes. However, not everyone appreciated the shift to including video games in movies. Canby moaned about the intellectual depth of *WarGames*, stating that it was "an entertaining movie that, like a video game once played, tends to disappear from one's memory bank as soon as it's finished."[33]

WarGames explored growing anxiety over computer hacking and the possible infiltration of secure corporate and governmental computer systems. In 1982, a group called the 414s targeted a range of computer systems, including the mainframe of the Los Alamos National Laboratory. The 414s were later revealed to be six teenagers from Milwaukee, Wisconsin; their motivations were simply "curiosity" and "fun."[34] Concern grew over the potential for individuals to perpetuate serious acts of terrorism. The US Congress responded with new laws. Along with the 414s, the movie *WarGames* inspired an update to the Computer Fraud and Abuse Act (1984), with Republican Dan Glickman even showing a clip from the film during debate on the bill. Hacking culture crossed into mainstream social awareness. Meanwhile, the potential for hacking to fundamentally alter society was explored in William Gibson's cyberpunk novel *Neuromancer* (1984), in which Case, a failed computer hacker, is hired by a mysterious employer called Armitage to merge two powerful artificial intelligence entities.

WarGames the movie quickly spawned *WarGames* the game. A movie about computers appeared ideal for conversion. Purchasing the MGM license, home electronics company Coleco anticipated gamers would likely relish the opportunity to play against the computer featured in the Hollywood blockbuster. Video game adaptations for the ColecoVision console, Atari 800, and Commodore 64 personal computers followed the movie release. ColecoVision programmer Joseph Angiolillo designed the game world. In the game, the computer mimicked Joshua, with the player assuming the role of program architect Professor Falken. Angiolillo replicated one specific scene from the movie, the NORAD face off, and used it as a game level. With little time to react, the player faced incoming missiles heading toward a map of the United States. DEFCON status displayed on screen for six regions of the country. The aim of the game was to stop any nuclear missiles from reaching their targets. The successful player navigated continual screen swapping in order to home in on each of the six regions to eliminate the enemy. Pressure grew as the torrent of jets, submarines, ICBMs, and satellite particle beams increased. Players faced a choice of endings: a ceasefire leading to diplomatic resolution, a counterstrike, or NORAD engaged.[35]

WarGames was not the first adaptation of a movie title for a video game. However, the release of the Coleco title in 1984 coincided with a spate of movie-to-game conversions that together indicated a fresh synergy between media formats. Partly a corporate mechanism, the movie-to-game crossover pointed to early forms of convergence, a term associated with media scholar Henry Jenkins.[36] In 1978, *Superman* was licensed for release on the fledgling Atari console. In 1982–1983, the movies *Raiders of the Lost Ark* (1981), *Alien* (1979), *Texas Chain Saw Massacre* (1974), *Halloween* (1978), and *The Empire Strikes Back* (1980) all became Atari VCS games. Despite the sense at the time that gaming culture primarily targeted kids, the inclusion of several R-rated horror movies intimated that an adult market also existed. By 1982, 10 million Atari consoles had been sold. Its high price tag clearly suggested a significant number were being used in adult-only households. The growing filmic influence also highlighted video games' ongoing search for identity. Among the various types of media, film was the closest to video games and continued to be a source of inspiration. Part of a broader Hollywood exploration of Cold War and nuclear fears, *WarGames*—the movie and the game—sat alongside other nuclear thrillers, including *The Day After*, *Testament*, and *Red Dawn*. Gamic perception of nuclear war drew on these wider filmic resources.[37]

WarGames coincided with President Ronald Reagan's public announcement of the Strategic Defense Initiative in 1983. The game involved the player in a fictitious SDI scenario at roughly the same time the public heard about scientific progress to make SDI real. Again, video games provided a critical take on Cold War proceedings. Like *Missile Command*, *WarGames* avoided identifying the aggressor and articulated a pacifist line. Neither the manual nor the software featured any mention of the Soviet Union as the enemy.[38] Instead the player saw an anonymous, computer-based adversary attacking a range of targets that included missile silos, Strategic Air Command bases, and cities. *War-Games* presented nuclear war as practically unwinnable, with the very best scenario being avoidance. In both the movie and the game, Joshua's conclusion that "the only winning move is not to play" echoed the sentiments of *Missile Command*.

The blurring of simulation and reality central to the story of the movie also framed the gamic experience. Like *Missile Command*, *WarGames* created the illusion of the player being a missile commander at a nuclear control station. The Coleco game manual, labelled as a "defense manual," informed the player that "you are the commander at NORAD defending the United States against enemy attack." It then set out detailed instructions on how to avert Armageddon. The illusion relied on the simulation of a realistic nuclear scenario; the replication of defense computer screens and emergency alerts fostered fear and conveyed the pressure of the situation. The Coleco manual continued, "You hold the fate of the United States—possibly the world—in your hands. Can you save the world from Doomsday?" However, no Coleco staff knew the actual architecture of the NORAD command center nor how national defense codes operated. The digital recreation of NORAD instead drew on images from the movie, which in turn were the work of Hollywood set designers (the NORAD set for the film version of *War-Games* cost over $1 million). Simulacra built on simulacra. *WarGames* existed as an invented world, but one with a sense of realism forged by the movie. The confusion over war game, simulation, and real life highlighted the potential of games to blur sensory experiences and collapse boundaries. Coleco felt the need to explain to its audience that "*War-Games* is a simulation of a nuclear strike. Some terms used here are different from those actually used by NORAD." Meanwhile, advertising for the game recycled the tagline for the movie: "Is it a game, or is it real?"[39]

The Cold War lens of *WarGames* also shone light on the shifting relationship between American society and digital technology. Coleco's title explored increasing reliance on computers and the potential pitfalls that this entailed. Hacked defense computers could destroy the world, and artificial intelligence posed a fundamental risk to American society. However, *Wargames* equally idealized the computer whiz and celebrated the hacker. Carly Kocurek claims that Badman's movie helped cast "the boy genius/hacker as desirable, a potentially powerful fighter in the Cold War race of arms and ideas" and, alongside Disney's *Tron*, promoted a new technomasculinity based on digital intelligence and the rejection of brute force. For Kocurek, *WarGames* offered "gamers as unlikely heroes."[40] Michael Newman notes how "early video games, like other emerging media, were an occasion for hopes and fears not only about the medium but perhaps more importantly about the society into which it was emerging."[41] In the case of *WarGames*, film and game articulated elements of both technophilia and technophobia. The game appeared rife with contradictions. The Coleco manual commanded, "Stop the NORAD computer from launching an ICBM counterstrike that will surely result in the destruction of all mankind."[42] At the same time, players had to *use* the NORAD computer to control planes and submarines, before they could tackle the computer itself. In gameplay terms, the player faced an intriguing prospect of both working with and against NORAD, or with and against the machine.

On a more abstract level, *WarGames* applied the concept of Murphy's Law (whatever can go wrong, will go wrong) to modern machine culture. With increasing reliance on complex computer systems, the specter of unanticipated consequences loomed large in 1980s America. In the original *Missile Command*, programmers provided no explanation for attack. In *WarGames*, the inability of computer and player to navigate the boundaries of real and fiction invited disaster. Joshua, the computer, is expected to "know" the difference between real and simulation, but something goes wrong. In real-life nuclear America, combined computer failure and human error had already produced disastrous results. The partial meltdown of the Three Mile Island nuclear plant in Pennsylvania in March 1979, owing to a faulty pressure relief valve coupled with maintenance oversight, highlighted the fallibility of both machine and overseer. In September 1983, around the time of *WarGames'* release, Stanislav Petrov, a Soviet controller, recognized just in time a false reading on a Russian satellite readout that showed a US attack, thus narrowly averting global nuclear war. The real NORAD computer system at Cheyenne Mountain also had its flaws. Twice in June 1980, faults in computer communications led to "nuclear attack" messages being relayed to US Air Force command posts. Doubts grew over the practice of relying on computers for the world's defense. Back in the 1960s, the science fiction show *Star Trek* had extrapolated where such technological dependency might lead. The 1967 episode "A Taste of Armageddon" depicted a streamlined war between two planets simulated by computers, with machines calculating casualties and selecting those to die in disintegration chambers. Citizens dutifully followed the command of technology, but the planets remained completely unharmed. Always the maverick and appalled by the situation, Captain James T. Kirk intervenes and destroys one planet's war game computer, forcing both communities to either adjust to peace or engage in traditional conflict that would ultimately destroy whole regions and ecologies. As Kirk highlighted, "actual war is a very messy business."[43]

Along with Murphy's Law, video games, reflecting broader popular culture, explored a growing fear of the sentience of machines. The video game version of *WarGames* began with the line "Greetings, Professor Falken," alluding to a computer capable of conversation, independent thought, and even action. Such fictions predicted a fundamental shift in the balance of power between humans and technology. Titles such as *WarGames* raised deeper philosophical questions of who or what was in control of life. Had humans gone too far in their engineering? The Frankenstein myth returned in pixel form. *Wargames* highlighted a cultural turn toward a fear of machines throwing spanners in the works (ironically explored through a computer game). Fear of computers became a popular theme of the period. James Cameron's early film *The Terminator* (1984) took the ideas within *WarGames* one step further by giving the machine ultimate power. The movie documented a failed war against the machines, after defense system SKYNET becomes self-aware and unleashes nuclear weapons on the world.

WarGames was not the only Cold War title released on the Coleco-Vision platform. In *War Room* (1983), players navigated dangerous spies and explosive satellites in a face-off between Russia and America.[44] Cold War strategy titles appeared on most gaming platforms. Sid Meier's *NATO Commander* (1983) was available for 8-bit home computers, while geopolitical title *Balance of Power* (1985) by Chris Crawford was popular on the Apple Macintosh. Marketed for a range of home computers including the Commodore 64, *Theatre Europe* (1985) drew protest from the Campaign for Nuclear Disarmament (CND) in the United Kingdom for its offensive content. Games reinforced a popular sense of the Cold War as based around a global map and as a tactical war game that could only be solved, via strategic oversight, computer assistance and comprehensive knowledge of the spheres of influence. Games thus pandered to a very specific reading of the Cold War. Intriguingly, they intimately involved players in the big decisions of conflict, but also shielded them from the personal repercussions or grim realities of such actions. Nuclear conflict was never something close or deadly, but distant and symbolic. War remained very much a war game.

Satire, Film, and Comedy in Nuclear War

In 1989 New World Computing, a small California programming house known for Might & Magic role-playing titles, released the strategy game *Nuclear War* for the Commodore Amiga and IBM PC. Like other war-based titles before it, *Nuclear War* envisioned Cold War conflict as a turn-based strategy game. Players competed against a group of computer-controlled world leaders, all jousting for global dominance and deploying nuclear weapons as their instruments of influence. Traditionally, war games offered realistic tactical simulations of conflict. New World Computing diverged from this norm by presenting war as a lively, satirical, and humorous affair. Programmers produced a game rich in satire of Cold War culture. World leaders appeared on screen as cartoonish characters with pun names, making cheap jibes at each other. Sometimes strange cow-missiles fell on countries.[45]

New World Computing programmers borrowed heavily from existing media to fashion their satirical take on the nuclear arms race. In 1965, Doug Malewicki designed the card game *Nuclear War*, a humorous strategy game detangling core elements of the arms race and involving players in such issues as "top secrets," "beatnik pacifists," "population explosions," "megaton bombs," and "drunken ambassadors." Advertised as "a comical cataclysmic card game of global destruction," Malewicki's game attracted a cult following. New World Computing programmers replicated the same kind of humor with their video game.[46]

The film *Dr Strangelove, Or: How I Learned to Stop Worrying and Love the Bomb* (1964) similarly influenced game design. Stanley Kubrick's movie was well-known for its parody of 1960s nuclear politics. Game sequences served as homages to the seminal anti-nuclear movie. New World Computing introduced *Nuclear War* with a B52 bomber emerging from the clouds, ridden by a military cowboy, dropping an atomic weapon onto a target below, with a mushroom cloud clearing to reveal the game title. Albeit in rudimentary, cartoon graphics, the sequence replicated the denouement of Kubrick's movie, when Buck the cowboy rides a missile down to its Soviet target, thus heralding nuclear war. The copycat scene revealed that the programmers were keen to place their game within the same narrative frame of the movie. New World Computing hoped for a gamic *Strangelove*. Similarities went beyond the credit screens. The obvious madness of General Jack D. Ripper in Kubrick's movie, who ordered a first strike on the Soviet Union with the sensationally ludicrous reason that the Reds had already polluted the "precious bodily fluids" of Americans, set the tone for *Nuclear War*. In the game, characters uttered bizarre phrases, offered unhinged facial expressions, and acted with the same kind of total irresponsibility as characters in Kubrick's movie did. Both New World Computing and Kubrick built comedy from absurd names, situations, and visual scenes. *Dr. Strangelove* featured characters such as Jack D. Ripper and Buck Turgidson, while the roster of *Nuclear War* included Ronnie Raygun and Gorbachef. The inane moments of *Strangelove* inspired equal bizarreness in the digital atomic world, where nuclear technology interchanged with bovine technology and aliens popped up onscreen.[47]

Cold War technology appeared in both *Dr. Strangelove* and *Nuclear War* in the guise of missiles and machines. Early 1960s-style computers —their noises, movements, and scale—featured prominently in both film and game. An early scene in *Dr. Strangelove* entailed Captain Mandrake checking printouts from huge UNIVAC defense computers at Burpelson Air Force Base. Mandrake is dwarfed by the computers' physical and mechanical dominance. Clearly, a technological frame guided the viewer, with the atomic world conceived as a secret realm of calculations, code words, and computer chips. Once set on its course, technology seemed destined to trigger destruction. The launch code CRM114, responsible for World War III in the movie, highlighted the role of computer information as crucial to nuclear conflict. Similarly, *Nuclear War* forwarded technology as the facilitator, even the partner, of Armageddon. The view screen (dubbed the "diplomacy terminal") replicated a defense computer terminal, with the player navigating a series of windows and buttons. The player viewed war through the display of the machine. Computers provided the crucial mediating force. War also seemed a game reliant on computers for their mathematics, as the player was encouraged to think in terms of numbers of megatons, population figures, and material resources.

Neither film nor game identified computers as the root cause of conflict. Unlike *WarGames* and its articulation of machine error, both *Dr. Strangelove* and *Nuclear War* focused on human fallibility and personal situations unfolding out of control. The madness of man, not machine, proved pivotal to the narrative. As Mandrake pointed out in the film, "the most likely way for war to start nowadays is by an accident, or a mistake, or by some mentally unbalanced person." Sidelined in both productions, the rational person had little say in events, as when Mandrake offered, "We don't want to start a nuclear war unless we really have to, do we?" to no reply.[48] Millions of lives appeared expendable in both film and game worlds, with Ripper unfazed by the loss of 60 million lives and *Nuclear War* announcing losses of millions by the minute. However, a crucial divergence emerged in assigning which group of people were to blame. In *Dr. Strangelove*, Ripper suggested a wider sense of guilt for doomsday when noting that "war is too important to be left to politicians." In *Nuclear War*, world leaders obsessively dictated the course of warfare and rushed toward oblivion. With ticker-tape headlines such as "Tricky Dick delivers death to your country," politicians clearly made the apocalypse happen.

Along with taking its cues from *Dr. Strangelove*, New World Computing's title reflected satire typical of the 1980s Cold War. Comic interpretations of atomic weapons and impending disaster were popular in the period. Black comedy served as one emotional coping mechanism of the period. Dark humor offset nuclear fear. Released in 1982 as a protest documentary, the movie *Atomic Café* spliced together a range of vintage pieces of nuclear propaganda to highlight the incredulity, inaccuracy, and, at times, sheer comedy of government broadcasts of the nuclear age.[49] *Atomic Café* offered sophisticated satire based on a careful selection of evidence. The Albert Pyun movie *Radioactive Dreams* (1986) superimposed noir detective fiction, swing music, and comedy onto a post–nuclear holocaust world.

Nuclear War seemed closest in tone to the cartoon puppetry of the British adult comedy series *Spitting Image*, which provided a humorous exploration of contemporary affairs using rubberized caricatures of leading figures. Reagan and Thatcher were regular "guests." A *Spitting Image* special entitled *The Ronnie and Nancy Show* aired on NBC in 1987. The rock band Genesis used the Reagan and Thatcher puppets in the music video for "Land of Confusion" (1986), in which Reagan wakes from a nightmare and presses the nuclear red button by accident (confusing "nuke" for "nurse"). *Nuclear War*'s leaders closely resembled these cartoon characters in looks and behavior. As with *Spitting Image*, the game poked fun at politicians and governmental authority.

Whether by choice or programming limitations, leaders in the game all acted curiously alike. Regardless of background, every politician appeared untrustworthy, irrational, and hell-bent on conflict. New World Computing resisted taking a side in the real Cold War conflict by presenting everyone as villainous. The game's programmers depicted America as one of five vague geographical and political entities on a strategic map. The American landscape appeared as a sizable green swath marked only by missiles, cities, and, eventually, atomic-carved craters. The player could face off against up to three American leaders in a game, each in charge of their own territories, thus leaving the United States open to attack by its own former presidents Nixon, Carter, and Reagan.

The notable presence of Ronald Reagan reflected his prominence in Cold War affairs and his responsibility, at least in part, for the escalation of global tensions in the 1980s. *Nuclear War* emerged from the protest culture of the period, in which Reagan represented an obvious target for attack. Reagan's in-game moniker Ronnie Raygun was already being used widely among US anti-nuclear groups, which combined the cartoonish image of Raygun with *Dr. Strangelove* imagery. For example, a 1981 issue of the newspaper produced by the California protest group Abalone Alliance carried a cover image of Raygun riding a nuclear weapon as a *Strangelove*-style cowboy.[50] *Nuclear War* regurgitated the same visual satire. It reflected familiar criticisms of the president as a gung-ho leader. The game manual stated, "[Ronnie Raygun] sometimes thinks that he is living in a Spaghetti Western, [which] makes Ronnie somewhat of a mindless warmonger. Ronnie is always a step away from his favorite, shiny toy, The Button."[51]

The inclusion of Jimi Farmer and Tricky Dicky seemed less infused by protest culture and more about including recognizable American characters. Programmers likely chose Nixon as an established popular caricature, with Carter included as the immediate predecessor to Reagan (but hardly suited as a warmonger given his work to orchestrate a détente). The game featured a decidedly US gaze on world affairs. Other leaders represented traditional enemies of American foreign policy, including Soviet premier Gorbachev and Cuban revolutionary "Infidel Castro." The game also included a character named Ghanji; having a character based on the world's most illustrious pacifist seemed an odd choice but demonstrated that the programmers happily placed humor and farce above serious comment or realism.[52]

Nuclear War recycled familiar criticisms of nuclear diplomacy. Such policies as mutually assured destruction informed the game world. People became numbers, meaningless casualties, with "Bomb shelters in use, only 11 million killed" a typical example of onscreen announcements. Gamers found it almost impossible to win at nuclear war. As each individual leader dropped out, their remaining nuclear arsenal automatically launched, unleashing a shocking impact on the world map. Most of the time, games ended with the earth exploding on screen or when one sole survivor was declared the victor, depicted as a madman jumping up and down on a nuked wasteland, shouting with a tone of desperation, "I've won, I've won." The game provided no spoils of war or gratifying sense of accomplishment. If the player lost, the game continued (albeit speeded up), with the other premiers still vying for world domination. As a helpless voyeur, the player watched the destruction continue on screen. New World thus gamified nuclear fear. It turned protester anxiety about the world's end into a game in which the player was trapped in the process. Thus, on a deeper level, *Nuclear War* offered a shrewd articulation of 1980s nuclear anxiety.[53]

Much of the appeal of the game stemmed from its ability to immerse the player in a chaotic and fast-paced drama. The speed at which other leaders took the offensive left the player with little time to relax. A news ticker on the bottom of the screen constantly updated casualty figures, with a typical announcement being "radioactive beta rays kill 100 million." *Nuclear War* jettisoned conventional game staples, such as high scores, in its pursuit of atmosphere and foreboding.

Nuclear War intimately connected the player with the nuclear technology of the era, producing a playful realism to the proceedings. Like other Cold War–themed games, *Nuclear War* encouraged the player to assume control of world events via a re-creation of a government computer terminal. The terminal provided a real-time window on events and the menu options to control them. "Welcome to Nuclear War. You have the Con," announced *Nuclear War* as players sat down for a session. Sometimes, players could see themselves taking action on the view screen. In particular, players launched weapons by pressing a button on their keyboard, which was animated as their thumb pressing a red button on screen. With little to differentiate onscreen and offscreen actions, connection to the digital realm was heightened. *Nuclear War* articulated a limited but nonetheless clear sense of cyborgization (a concept usefully explored by Donna Haraway).[54] Players saw the direct consequences of their actions, of their button pressing, by watching nuclear explosions on screen.

The digital act took on wider significance thanks to the mythology in atomic culture surrounding the red, or nuclear, button. Commonplace in nuclear documentary and Hollywood fiction, the red button served as a key symbol of the end of the world. Video game culture in the 1970s and 1980s included the red button, but granted it a subtly different meaning. At arcades, red buttons on machines represented action buttons to make things happen and win games. In wider computing circles, the red button represented a reset or a reboot. *Nuclear War* thus offered an intriguing confluence of computer-generated and Cold War–induced meanings.

The simplification of Cold War events into a sequence of hurling missiles at one another made for basic, repetitive gameplay. Players chose from just four options: build weapons, release propaganda, defend, or prepare to strike. No buttons existed for peace, diplomacy, or negotiation. No solution existed other than destruction. This reductive reading of conflict had some power and legitimacy thanks to the familiarity of the anti-nuclear message in the 1980s. As with other anti-nuclear fiction, *Nuclear War* depicted nuclear conflict as fundamentally about the exchange of missiles and the loss of cities. The distinctiveness of *Nuclear War* came from its total dedication to presenting atomic warfare as, first and foremost, a game. Announcements such as "your country builds nuclear toys" highlighted war as a game played by world leaders with little sense of real-world consequences. Cartoonish presentation underlined nuclear weapons as "toys" in a childish recreational activity.

While clearly a product of public fear, *Nuclear War* failed to be taken that seriously as a criticism of Cold War politics. Games took on controversial topics, yet their deeper messages continued to be ignored into the late 1980s. As one reviewer for game magazine *Zap!* wrote, "Ever fancied causing death and destruction across the world? Ah, who cares? Blast 'em anyway!"[55] Video games seemed far less successful than other media in promoting critical thought. US Gold, the publisher for *Nuclear War*, avoided any moral or political outrage concerning the title. Having previously released *Raid Over Moscow* (1984), a controversial Cold War title banned in Norway, US Gold shrewdly added a warning label to *Nuclear War* concerning its "offensive material." The offensive material included an awkward joke about the "Chernobyl effect" (referring to a meltdown in a player's country). A reviewer for *Amiga Power* magazine described *Nuclear War* as full of "utterly feeble puns and dubious taste."[56] Another reviewer interpreted the game as essentially harmless: "With over half a billion people dying per game, space aliens and flying cattle, it is hard to be offended by *Nuclear War*."[57] While *Missile Command* avoided real-world associations by using symbolic and abstract graphics, *Nuclear War* sidestepped the harsh realism of radioactive warfare through its cartoon effects and dark comedy. It was also curiously behind the times. With the Intermediate-Range Nuclear Forces Treaty of 1987 and growing economic collapse in the Soviet Union, the real Cold War had begun to thaw by the late 1980s. On December 3, 1989, President George W. Bush and Mikhail Gorbachev officially declared the Cold War over. The video game thus ended up being an early form of Cold War nostalgia.

More than literature or film, atomic-themed games in the 1980s captured the technological dramaturgy of the Cold War. Video games provided a level of immersion and interactivity lacking in other entertainment media and connected with the Cold War experience in novel ways. The sense of modern warfare as something innately technological, viewed by remote satellites and on computer screens, easily translated into the video game genre. In the guise of buttons, chips, joystick controls, and readouts, computer technology proved fundamental to real-world conflict. War functioned through a digital interface. Computer games replicated this interface in a domestic setting. Marked by computer-programmed battleground simulations, the Cold War seemed ideal video game fodder. The two sides of the Cold War suited simple, competitive game play. Games captured the imagery of the Cold War and broke it down into core essentials. The simple iconography of warfare—incoming missiles and threatened cities—was easily transplanted into game worlds. Games and war seemed made for each other.

Given the popular imagination of the Cold War as a remote and inaccessible event, video games also provided a sense of participation in a distant conflict. They brought the Cold War into the home. A conflict with few options for practical involvement, and mostly based on saber rattling and missile quotas, the Cold War had shut out citizens for decades. Nuclear conflict revolved around two national premiers pushing red buttons for mutually assured destruction. The Cold War was a war of elites. Through video games, ordinary citizens could feel involved and able to make their own decisions. Some games granted key decision-making powers to players. Others pointed to the futility of conflict, with the player moving from helpless voyeur to equally helpless participant.

The Cold War was an unreal war. According to Guy Oakes, it was the definitive "imaginary war."[58] It existed most of the time as a simulation, a politically framed drama, and a dark fantasy. Much of the real Cold War was played out in war games and political–military simulation. Few people actually experienced the conflict. It "lived" in military computers. Games provided the opportunity to play out the dark fantasy and experience Armageddon. They were a natural extension of Cold War imaginings, providing furtive steps in a greater visualization process. Gamers imagined America under nuclear attack and duly defended their nation.

Cultural artifacts of the 1980s Cold War, computer games such as *Missile Command* and *Nuclear War* played one more important role: they functioned as protest. They recoded nuclear fear as gameplay. They transformed the grim and cloudy specter of nuclear conflict into a tightly programmed digital doomsday scenario. With their focus on missile defense systems, computer consoles, and binary digits, video games presented a technological reading of Cold War that contrasted heavily with anti-nuclear fiction of the time, such as *Testament* and *The Day After*, movies that prioritized the American family and emotional reaction. However, atomic games still captured the darkness of the period in their minimalist game over/end screens, dismal death tolls, and clever inversion of "winning" strategies. By their interactive nature, games presented the American public with a rare opportunity to directly face the Cold War going hot and the desperation that entailed. They showed the unwinnable nature of conflict. When the actual Cold War thawed in the 1990s, and superpower rivalry diminished, the atomic game lost its cultural relevance.

In the absence of the Soviet menace, US politicians, military leaders, and cultural commentators searched for fresh villains, despots, and disaster scenarios. While popular cinema cast European madmen as nemeses, the new adversaries in video games included Nazi agents in Dreamwork's *Medal of Honor* (1999) and hellish otherworldly creatures in id Software's *Doom* (1993). Only Westwood Studio's *Command & Conquer* (1995) kept the Soviet threat alive in gaming via a twisted sense of time and history akin to a Philip K. Dick novel. Most games that explored the Cold War increasingly did so through the lens of nostalgia, as in the *Fallout* series (1997–2015), first created by Interplay Entertainment and Black Isle, then later programmed by Bethesda. A first-person, open-world game, Bethesda's *Fallout 4* (2015) projected a 1950s world of bunkers, bombs, and shelter signs into a futuristic apocalyptic setting.[59] Bethesda indulged in all kinds of atomic nostalgia in the role-play title, from authentic kitsch songs of the 1950s playing in the background to a fictitious in-game religious cult, the Church of the Children of the Atom, worshiping "the glow." In a sense, *Fallout 4* itself functioned as its own church of atomic worship. Only fragments of the United States remained, although consumer life remained buoyant in the form of trading stores and bottle-cap currency (most bottle-caps came from Nuka-Cola soft drinks, a pre-war beverage that continued its popularity into the apocalypse). The *Fallout* series as a whole highlighted the continuing allure of immersion in an apocalyptic fantasy. The fiction of living through World War III remained potent even as the nuclear threat receded. *Fallout* explored the fall of America due to destructive tendencies, both internally and from communist aggressors. By the 2000s, a new enemy had also emerged in games: Islamic terrorism.

4
9/11 Code

A POLICE CAR races across Manhattan; a plane crashes into two sky-scrapers, creating a horrific fireball; water forms on the face of the Statue of Liberty, resembling a tear. The opening sequence of *Parasite Eve* (1998) by SquareSoft depicted a disaster scene now most identified with the terrorist attacks of September 11, 2001. Based on a 1995 Japanese horror novel by Hideaka Sena and programmed in California, the survival–horror title explored the idea of terror striking New York City.[1] In the first playable level of the game, lead character Aya Brea, a twenty-five-year-old New York Police Department officer, witnesses a mutant human called Eve cause havoc during an operatic performance at Carnegie Hall on Seventh Avenue. Eve creates a mass immolation among the audience. People panic and attempt to flee, while one burning man falls from the upper echelons of the theater in an act of sheer desperation. As emergency services arrive, one medic flippantly remarks, "Hope this becomes a TV movie," as he gazes at the unfolding tragedy. Mysteriously linked to the assailant, Brea sets out to save Manhattan by tracking Eve and confronting her at a variety of destinations, from Delacorte Theater to Central Park. A final scene plays out at Liberty Island, when the Statue of Liberty is broken. SquareSoft's *Parasite Eve* imagined a comprehensive terror attack on American soil, with New

York City destroyed by a surprise force. Set on Christmas Eve 1997 and released in 1998, the game predated 9/11 by three years.

Coding 9/11

The terrorist attacks on September 11, 2001, aligned the dawn of a new millennium with a provocative end-of-world scenario. As buildings fell, smoke gathered, and fires roared, a new era was marked by a monumental catastrophe. The terrorist attacks of 9/11 were shocking, devastating, and palpably real. They were marked by decidedly physical, visceral, emotional, and traumatic experiences with huge human costs. Almost 3,000 people lost their lives.[2] The ramifications of the event, in terms of the ensuing War on Terror, global terror attacks, and the toxic fallout of debris at Ground Zero, cast a long shadow across the world.

The Islamic terrorists who took control of four civilian aircraft on the morning of September 11 were inspired to act by a range of factors: the rise of religious fundamentalism, the leadership of Osama Bin Laden, and unwelcome American involvement in the Middle East. On a more subtle and symbolic level, the terrorism flowed from the growing forces of globalization and Americanization, allied with a sense that the United States was colonizing world cultures and values. Along with the "high gloss" of modernity, the "blunt force" of foreign policy, and the "perceived godlessness" of the United States, author Don DeLillo suggested, "It was the power of American culture to penetrate every wall, home, life and mind" that invited terrorist action.[3] With Western values and concepts, video games were no doubt part of that penetration: exporting Americanness in digital code.

Alongside the real and tangible costs—the lives lost and destroyed, the cracked concrete and twisted steel debris of the Twin Towers—the terrorist attack was deeply symbolic. It challenged the westward shift of world religion, politics, economics, and popular culture. It stood as an act of open defiance against the spread of American values. The terrorists chose the Pentagon as the prime symbol of US military power and the World Trade Center as the premier emblem of Western capitalism. They launched a Holy War against infidels. They challenged American cultural dominance by shooting down its icons. The terrorist attack was a bold and brazen attack on the West and its symbols.[4]

Terrorists used images to attack images. The attacks on 9/11 were part of a global image war and an unfolding iconographic conflict. The terrorists employed the same technology as US multimedia giants and corporations, but to broadcast an alternative image of America. Television channels around the world switched from broadcasting *Friends* and scenes in Central Perk to reporting on enemies and a fallen World Trade Center. Visuals depicting the rise of America were replaced with images of its fall. Al Qaeda exploited the same wave of visual culture that helped elevate the United States to cultural and technological dominance by using 24/7 media and fast-moving journalism to broadcast their actions to a world audience. The attacks on 9/11 were a cleverly coded event for the digital era, designed for multiple camera shots and instant reportage. Terrorists chose their targets with the media in mind; as Thomas Stubblefield noted, "the attack was aimed at and made for the image."[5] Neal Gabler observed that the hijackers were "creating not just terror; they were creating images."[6] The terrorists crafted a collage of disaster imagery, both shocking and numbing. According to Jacques Derrida, the terrorists sought to "spectacularise the event" for "maximum media coverage."[7] Deliberation over targets and timings reflected not just a well-calculated event, but an event with the media in mind; the second attack was timed to take place twenty minutes after the first, with the camera lens already focused on the World Trade Center for maximum impact. As Gabler explains, "it was terrorism with the audience in mind," while Paul Virilio labelled the event "televisual terrorism."[8] The attacks of 9/11 were designed to be viewed.

Those caught in the maelstrom of a struck Manhattan instinctively turned to digital devices to process the event. Cameras and phones provided a lens on the action. A safe distance from Ground Zero, millions watched images of New York City under attack through their television screens. With the scenes on screen not that dissimilar from a blockbuster disaster film, comparisons of 9/11 to a Hollywood movie quickly emerged. As Reza Aslan observed, "the events were truly *like a movie*: the hijacked planes, the crumbling skyscrapers, the crush of people on the ground suddenly shrouded by a cloud of ash and rubble. It all seemed as though it were plucked from a Hollywood script."[9] The similarity between 9/11 imagery and fictional catastrophe meant inevitable comprehension through the lens of film. The attacks on 9/11 took on an unreal quality. As if caught inside a movie scene, people experienced 9/11 as a series of "images as much or more than actual events."[10] The boundaries between movie fiction and terrorist reality blurred. Director Robert Altman even blamed Hollywood for inviting 9/11, arguing, "The movies set the pattern . . . Nobody would have thought to commit an atrocity like that unless they'd seen it in a movie."[11] That Hollywood thrived on repetitions of disaster imagery potentially helped craft both the terrorist act and the inducement of mass fear. For Derrida, despite "the killing of thousands of people, the real 'terror' consisted of and, in fact, began by exposing and exploiting . . . the image of this terror by the target itself."[12]

Visual culture imagined 9/11 before September 11, 2001. That visual culture included games. The "image/event" of 9/11 drew on established frames of reference and influences, including those from the digital realm. While film analogies dominated initial popular analysis, the images of terrorist attack resembled scenes from blockbuster video games just as much as those from blockbuster movies. Thousands, if not millions, of players had fought terrorists in virtual American worlds prior to 9/11, witnessing the destruction of US cities on their game screens. Popular series, including *Metal Gear Solid* and *Desert Strike*, had explored 9/11-type scenarios. September 11 was a gamic as much as a filmic event.[13]

The most disastrous of disaster movies, the worst imaginable level of a game, 9/11 represented an image/event of immense power. The power of 9/11 owed much to the intertwining of the event with an American catastrophe culture already widely fictionalized in film and games. Like an ascendant mushroom cloud, bringing all the debris into the atmosphere, 9/11 threw a kaleidoscope of pre-existing images into the air. For W. J. T. Mitchell, "the violent spectacle, the image of destruction," was all consuming.[14] Cast as the ultimate image/event, Jean Baudrillard declared September 11 a powerful radicalizing moment. Millions became transfixed by the catastrophic imagery. As Baudrillard explained, "the fascination with the attack is primarily a fascination with image."[15]

Three distinct images initially came to the fore after 9/11. In the first few days of constant twenty-four-hour attention, they became synonymous with the attack. For many watchers, they *were* the attack. The much-repeated video image of planes hitting the Twin Towers, with smoke engulfing the skyline and the streets below, dominated television screens; the static photograph of the falling man jumping from the North Tower featured on newspaper pages; and the New York firefighters holding a flag aloft amid the rubble served as a powerful marker of survival. Each image told a different story. Each held intrinsic visual power. Once seen, they proved difficult to remove from the retina or memory. Baudrillard noted the "unforgettable incandescence of the images" of 9/11.[16] Gradually, a meta-story about the terrorist attack emerged from these and other pictures. A greater artistic representation evolved, as mass media, film, literature, and games all processed the event.

Deletion / Twin Towers

From 1973 to 2001, the Twin Towers of the World Trade Center existed as real, concrete-and-metal superstructures, part of a tightly constructed urban landscape navigated by thousands of people daily. They were routine, ritualistic, and intimately connected to the twenty-four-hour business of the Big Apple. People sold shares, sent emails, and ate in cafes inside them. The towers also operated as deeply symbolic structures. Consummate engineering feats, they carried the weight of America on their girders. Once the world's tallest buildings, they pointed the way to the heavens of manifest destiny. They denoted the rise of twentieth-century business, capitalism, and credit. Baudrillard noted the "competitive verticality" of Manhattan skyscrapers, jostling for position in a global economic game.[17] The towers represented US superiority in that game.

The destruction of the towers symbolized the destruction of the United States by terrorism and the end of its superpower hegemony. Mitchell declared the implosion of the Twin Towers "the most memorable image of the twenty-first century" and an unparalleled symbolic event.[18] For Baudrillard, the towers represented twenty-first-century religious crosses, sacrificed as part of a religious spectacle.[19] For Virilio, media attention on the collapse of the World Trade Center (as opposed to the Pentagon, which was also attacked on 9/11) highlighted the skyscrapers as the crucible of US-style capitalism made frail.[20]

The footage of the Twin Towers falling provided the most powerful of televisual imagery. Capturing the precise moment of the terrorist attack, media stations broadcast views of the skyscrapers being hit and crumbling to the ground. The explosive imagery went viral. Footage of monumental collapse played continually across multiple media platforms. The destruction of the skyscrapers became the iconic portrait of the event. Universally distributed and universally consumed, the scene of implosion became the quintessential image of 9/11. Like the Zapruder home movie showing the assassination of President Kennedy in 1963 or the camcorder video of Rodney King being beaten in 1991, film had graphically captured the unfolding disaster and immortalized the moment of the terrorist attack.[21]

Part of the power of the image lay in its endless recycling. Twenty-four-hour news coverage played the short clips over and over again, frame by frame. The perpetual repetition of the image of the Twin Towers in flames became indelibly burnt on the television screen and on the retina of the viewer. Technologists note that when a static image remains on a LCD or plasma screen for a prolonged time, it leaves a ghost image on the screen similar to the screen-burn found on old cathode ray tube televisions, a case of image persistence. The destruction of the Twin Towers left incontrovertible image persistence. Entranced by the spectacle of disaster, viewers became visually glued to the image, unable to look away. Karen Engle refers to media coverage of 9/11 as forging "repetition compulsion."[22] Continually witnessing the moment of impact affirmed a sense of growing nationwide trauma. People shared the pain and horror by repeatedly watching the disaster play out. Geographic boundaries collapsed and empathy increased with every viewing. Repetition created a sense of being there—and of not being able to escape.

The repetition of the disaster reel made the event uncomfortably vivid and real. It was also filmic and gamic. The footage resembled a blockbuster movie. It repeated the scenario over and over like the stage of a game. Repetition tied 9/11 to gaming culture. Since the ascendancy of arcades in the 1970s, repetition remained a hallmark of game design. Video games operated as a succession of almost identical image/events, game screens replicating themselves with only slight variations of enemy placements. Players learned how to conquer foes by replaying levels, which allowed them to become familiar with the same drama and master the scene. Like these computer games, footage of 9/11 was repeated over and over on a screen. Like facing enemies in *Space Invaders*, people interfaced with 9/11 through a process of repetition and constant exposure. They interacted with 9/11 as a "repeat event." Power resided in this interactive process. Riccardo Fassone describes the "looping play experience" of games and their linkages to events like 9/11.[23] Both maintain an almost hypnotic influence over the viewer, with a visual transfixion based around duplication. A flow-like element takes hold in the process of continual exposure. Absorbed by the drama, people become caught in a loop, replaying the same stage over and over again. They play and replay, they watch and re-watch. Like an unwinnable level in a video game, 9/11 had audiences caught in an unresolvable scenario. They thus became collective victims of a repeated trauma.

The fall of the Twin Towers had other gamic associations. The buildings themselves featured in a range of video game worlds prior to 9/11, the towers existing in the digital realm just a decade after their physical rise. The original design schematics for the World Trade Center received criticism in the 1960s for their nondescript elements as well as for their damaging effect on small businesses. Judged architecturally inferior to other Manhattan icons, towers A and B were derisively likened to the "boxes that the Empire State Building and Chrysler Building came in."[24] The designs lacked surface detail or distinctiveness.

The aesthetic simplicity of the towers made them ideal for early video games. Simple (polygon-like) structures suited two-dimensional and three-dimensional games. Alongside the more challenging task of translating the Statue of Liberty into block pixels, game designers harnessed the simple lines of the World Trade Center as a visual marker of an urban skyline and city streets, with the two skyscrapers providing easy to fashion symbols of New York and the wider United States. The World Trade Center grew to have a digital life alongside its real one.

New York City (1984), created by Synapse for Atari 8-bit computers, featured one of the earliest pixel versions of the World Trade Center. Synapse programmer Russ Segal created a game in which the player, as a sightseer in the Big Apple, set about visiting twelve locations, including Grant's Tomb, the United Nations headquarters, the Empire State Building, and the World Trade Center. The game offered an impressively detailed recreation of "tourist New York." An overhead view of Manhattan Island marked major sights on a city grid. Like any tourist, the gamer faced the core challenge of mastering the metropolis by navigating its subway lines and road system. Players had daily tasks based around resource gathering, from managing car fuel to sating their hunger and, somewhat bizarrely, collecting animals from Central Park Zoo. The digital tourist drove around the city by automobile. When the player's energy depleted, an onscreen reminder ("time to eat") urged the player to locate the nearest Mart for burgers and pretzels. This early digital Manhattan seemed mostly unfriendly and a staunchly competitive place where cars blocked your way and people stopped you from obtaining food. The cityscape amounted to an assembly of obstacles. Survival for seven days (in game time) presented a simple but highly taxing aim. In 1986, the software label Americana, owned by US Gold, released the title in Europe, complete with a new action-packed cover that depicted a cartoonish character in the foreground of a skyscraper-dominated skyline. US Gold publicity read, "This hurried jungle of steel, concrete and glass bristles with fascinating sights and more than its share of danger. From the placid greenery of Central Park to the massive presence of the World Trade Center there is no city in the world like New York."[25] Sitting alongside other Manhattan structures, the pixel version of the World Trade Center existed as part of wider setting—part of a living, breathing cityscape (or as much of one as 8-bit technology could manage). Pre-9/11 digital New York appeared as a space for tourism, travel, and sightseeing, with the Twin Towers one of several sights to behold.

Like the rest of popular culture, video games asserted the World Trade Center as an iconic feature of the American landscape. The Twin Towers appeared in static backgrounds for a range of two-dimensional fighting games in the 1990s. *Streets of Rage* (1991), *Mortal Kombat 3* (1995), and *Tekken 2* (1995) all featured Manhattan backdrops with the World Trade Center clearly visible. In the 3D beat-em-up game *Fighting Force* (1997) by Core Design, New York City was a playable level, with the World Trade Center beautifully lit by office lights and a sunset in the distance. Programmers saw the Twin Towers as a familiar icon that players would instantly recognize on screen. A pixelated World Trade Center provided "visual placement" in a digital environment. Its presence instantly informed players of their whereabouts. Two-dimensional fighting games typically employed "eye-candy landscapes": gamers played consecutive bouts in order to open new levels, which allowed them to see all of the prerendered settings and fully explore the game.[26] The World Trade Center functioned as eye candy: it provided a visual treat, as well as something to admire during breaks between pummeling your opponent into submission. Video games cast the Twin Towers as a welcome discovery. The landscape portrait–style of games granted the pixelated World Trade Center a static and frozen-in-time quality. Contenders came and went in the foreground, but behind them the Twin Towers stayed true like a mountain range. They exuded a sense of permanence and indestructibility.

Racing games treated the World Trade Center in a similar way by commandeering the towers to give a sense of place, beauty, and grounding. They also granted the buildings a minor ludic role. The towers operated as a landmark to get one's bearings by and an obstacle to navigate. *Gran Turismo 4* (2004), *Midnight Club* (2000), *Crazy Taxi 2* (2001), and *Driver* (1999) all featured the World Trade Center. In *Driver*, designed by Reflections Interactive, gamers played as Tanner, an undercover NYPD detective with peerless driving skills, with the World Trade Center visible in the game world. In Sega's *Crazy Taxi 2*, players ferried visitors around Manhattan in yellow cabs, sharing with their occupants the delights of the Big Apple.[27] Like the 8-bit *New York City*, Sega players navigated digital New York primarily as a tourist attraction. Gamic New York resembled a vacation advertisement. The pixel Twin Towers not only identified the skyline as Manhattan to the player, but symbolized a highly desirable and exportable American way of life. Pre-9/11 titles presented New York as fundamentally a fun destination. Rather than anchorless and abstract, the digital landscape, with its landmarks like the World Trade Center and the Statue of Liberty, seemed rooted in material life.

Along with fighting and driving activities, game designers connected the Twin Towers to flying. The grounded and stationary landmark became tied to speed, piloting, and air travel. A number of pre-9/11 games presented the Twin Towers as a structure to navigate: sometimes to fly by, other times to destroy. In Nintendo's *Pilotwings 64* (1996), players faced a variety of increasingly taxing challenges based on flying across a range of miniature worlds. Nintendo's cartoon version of America, entitled Little States, featured a blocky World Trade Center alongside the Statue of Liberty. In the New York Metropolis Dance level, the player, aided by a jetpack, was tasked with flying through a series of nine rings in the city sky.[28]

The idea of New York skyscrapers as something to destroy in flying games first appeared in home computer programming code. In the late 1970s and early 1980s, game magazines routinely published programs that readers could type into their machines. Inputting around a hundred or so lines of BASIC code resulted in simple, playable games. In December 1979, *Personal Computer World* magazine published the code for *Air Attack*, a rudimentary bombing game for the Commodore PET. Based in the United Kingdom, Simon Taylor programmed a graphically improved version of the title in September 1981 for the Commodore Vic 20 and set it in Manhattan. Taylor, who had never been to New York City, called the game *Vic New York*; as he explained in an interview, "Where else has lots of skyscrapers?"[29] Taylor initially promoted the title himself, "selling the games by recording them to cassette, getting on trains and buses and selling them to local computer shops," before gaining a contract with Commodore in April 1982. Commodore retitled his game *Blitz*. In 1984, budget range Mastertronic sold the game as *New York Blitz* for $1.99. Taylor's game roughly followed the conventions laid down by the original PET code: *Blitz* was a flying game in which the player had very little control over the flying. Rather than piloting a plane, the player watched from a distance as a civilian aircraft gradually descended toward a range of skyscrapers. Imagining being on board the flight, the gamer then dropped bombs to clear the ground below for a safe landing; as the instruction leaflet stated, "the only way you and your crew are going to land safely is if you flatten the city."[30] Pressing the keyboard's space bar released the explosives. The more astute players knocked out the tallest skyscrapers first, to avoid the plane hitting their sides. The challenge was then repeated with a variety of backdrops, starting in Miami and moving across the United States, with the aim to "fly up the East Coast to reach New York." Set in 2002, the final level required the player to clear New York City of all buildings, that is, to destroy Manhattan. Mastertronic's cover art depicted a plane heading directly toward New York skyscrapers, with a fresh explosion emanating from the side of one of the closest high-rise buildings. The game was appealing because of its accessibility and simplicity. As Taylor explained, "the tension is built because, like a real plane, it has to fly at a certain speed . . . then, it's a combination of timing, looking ahead." The game also required the player to place personal survival above the collective fate of city residents. While the abstract nature of the graphics disguised the grim reality of the task (the game showed no people dying), for some, it was a distasteful experience. Remembering the title, one gamer recounted, "I wasn't allowed to play it by my parents because of the plot."[31] However, for Taylor, as with many other gamers, "it was just an action puzzle game really. When the attack on the Twin Towers happened, I did feel a bit of a pang of guilt, but of course, I was never to know."

First released in November 1982, *Microsoft Flight Simulator* offered a more realistic sense of flying and the dangers involved than previous games did. An update of a wireframe Apple game, *Microsoft Flight Simulator* promised "a full-color, out-the-window flight display," including weather effects and time of day settings. One advertisement claimed, "If flying your IBM PC got any more realistic, you'd need a license."[32] Program updates and expansions strengthened the claim to realism. By the late 1990s, the simulation-based title featured dynamic scenery, real-world weather, GPS mapping, the latest civilian aircraft, and facsimiles of over 20,000 real-world airports. Microsoft also toyed with "visual damage" settings, leaving it on in some versions, while relying on programmer "mods" to engage it on others. In a direct and practical way, *Microsoft Flight Simulator* linked gaming to 9/11. Along with courses at a Florida flight school, Al Qaeda terrorists were rumored to have used the game to practice their skills on virtual Boeing jets to prepare for the September 11 attack.[33] While thousands of flight simulator pilots over the years saw the towers on screen and adjusted their dials to fly toward them (and even explode into the buildings), the terrorists did so with real-world intentions. The flight simulator helped people visualize the World Trade Center as a target and practice the attack scenario on repeat. The game provided a dress rehearsal and potent simulation of what was to follow.

Another early title to explore Manhattan under attack marked the birth of what may be the most famous video game character, Mario. In the Nintendo arcade game *Mario Bros.* (1983), the Brooklyn-based plumber defended his city from creatures exiting sewer pipes. Creator and Nintendo visionary Shigeru Miyamoto chose New York City as the setting because he imagined the underbelly of the city as a "labyrinthine subterranean network of sewage pipes."[34] That same year, Creative Software launched *Save New York* (1983) on the Commodore 64, pitting the player against an alien attack. In a game style reminiscent of *Space Invaders*, the player shot down aliens before they destroyed skyscrapers. Aliens literally consumed the Big Apple by eating buildings, with most structures in a state of imminent collapse. A reviewer for games magazine *Ahoy!* hypothesized, "We don't know for a fact that Joe Jetson, designer of this game, is any more interested than Jimmy Carter was in saving New York," referencing President Carter's neglect of city problems in the 1970s.[35] Rather, the US setting likely reflected the over-familiarity of Tokyo for Godzilla-style monster attacks; thus, "Creative Software might have deemed a change of locale necessary."

Game designers also imagined the destruction of the Twin Towers themselves. While the Statue of Liberty featured more often in digitally based disaster scenarios, several titles entertained the notion of a targetable and destroyable World Trade Center. In the smash-em-up *King of the Monsters 2* (1992) for the Neo Geo console, huge, colorful monsters stomped on the World Trade Center, along with other high-rise buildings. Inspired by King Kong and Godzilla, the arcade title *Rampage* (1986) by Bally Midway, required players to destroy a half dozen skyscrapers with their own monster. In 1994, Electronic Arts released *Urban Strike* (1994) for the Sega Genesis console, the third in a series of isometric helicopter action/strategy games with backdrops related to the Cold War, the Gulf War, and the (at that time fictious) War on Terror. In *Urban Strike*, one level detailed an unexpected attack on New York City by terrorist and cult leader H. R. Malone. While shooting down Malone's armed minions and disarming his bombs, the player witnessed a laser superweapon being used to destroy the World Trade Center. With the side of one tower missing and flames from the building rising up, the player attempted to rescue survivors by picking them up in a helicopter. The game scenario was likely inspired by the real-life terrorist bombing of the World Trade Center in February 1993, when a truck bomb with nitrate-hydrogen gas detonated in an underground parking garage. The *Strike* series employed a fictive timeline, in which the player combatted Malone's terrorists and their attack on the Twin Towers in the year 2001.[36]

In the late 1990s, visions of destruction in gamic New York City became more powerful and dramatic. Themes of terrorism, conspiracy, and technological and social breakdown marked the new urban template. A perceptibly different digital New York City emerged, one marked by impending catastrophe. Once a celebrated symbol of America, the city now appeared highly vulnerable. A number of titles explored a fundamentally darker and threatened United States through the lens of a besieged East coast metropolis. In the stealth games *Syphon Filter* (1999) and *Syphon Filter 2* (2000), developed by SCE Bend Studio, the player navigated both terrorist and government-inspired conspiracies, with first Washington, then New York, the focus of attention, while in *Command & Conquer: Red Alert 2* (2000), Soviet-led terrorists took over Manhattan.[37]

The game world of *Deus Ex* (2000), designed by Ion Storm, perhaps offered the most cogent and immersive of story lines. The title depicted a dystopic New York City in a state of terrorist control. It began with the player controlling J. C. Denton, a UNATCO (or United Nations) operative, on a mission to regain control of Liberty Island by eliminating National Secessionist Forces (NSF) terrorists (and liberating US citizens who had been taken prisoner). The setting was conspicuously dark, with both the head and the torch of the Statue of Liberty missing, and a dense fog engulfing the island. As with other dystopic fiction, Liberty Island took on the role of a visual indicator of the declining health of civilization. The scene was reminiscent of the denouement of Franklin Schaffner's *Planet of the Apes* (1968), in which astronaut George Taylor (played by Charlton Heston) discovered a partly buried Statue of Liberty on a deserted beach, providing incontrovertible evidence of the demise of humankind and the loss of democracy, marked by the rise of fascistic apes. *Deus Ex* similarly positioned a broken statue as a potent symbol of both democracy and civilization threatened. Programmers highlighted the danger to American values by pointing out, "Liberty is just a statue." A sense of symbolic terrorism pervaded the game world. Sheldon Pacotti, lead writer for *Deus Ex*, explained, "The game's most accurate prediction comes in how terrorism operates in the story. From the mission ideas to the way characters strategize and explain themselves, the game dramatizes the idea that 21st century war will be waged with symbolic acts."[38] The conspicuous absence of the Twin Towers on the skyline of *Deus Ex* was later read as a symbolic act in itself. Developers accounted for the noticeable gap by explaining, "We just said that the Towers had been destroyed too."[39] In an act akin to the beheading of the Statue of Liberty, the Towers crumbled to the ground, but off-screen and out of sight.

Deus Ex reflected growing fears of terrorism and conspiracy in 1990s America. The game challenged the player, as Denton, to navigate a realm of ever-changing allegiances, continual backstabbing, and underlying threat. The game existed in the same fictional terrain as the *X-Files* television series (1993+) and *X-Com* game series (1994+), both franchises pandering to Generation X fears of hidden cults, untrustworthy government officials, and the rise of a New World Order. By replacing clearly identified heroes and villains with a murky moral maze, *Deus Ex* undermined not just a cowboy-versus-Indian or American-versus-

Soviet view of national affairs, but also traditional binary concepts of play. According to Ian Bogost, this "procedural rhetoric of moral uncertainty" clashed with traditional conventions in games, in which a "moral system is assumed and enforced through a set of unit operations for Army procedure."[40] The pre-9/11 ambiguity of *Deus Ex* would soon be replaced with a more comfortable binary position, with titles following the 9/11 terrorist attack, such as Infinity Ward's *Call of Duty* (2003), pushing a simpler narrative.

The destruction of the Twin Towers in games also reflected a growing late-twentieth-century interest in an end-of-days scenario. While Mathias Nilges contends, "We have always been fascinated by cultural depictions of destruction," by the 1990s, millennial fears, instability brought on by the end of the Cold War, concern over technological reliance and machine sentience, and social unease heralded a new spike in fictional accounts of world disaster.[41] Alongside blockbuster films such as *Armageddon* (1998) and *Deep Impact* (1998), video games contributed to a broader cultural exploration of what disaster might mean for the United States. Games weaved together old threats (Soviets and spies) with new ones (terrorism and cults), and melded biological weapons with technological advancement, often with little attention to chronological realism. Most of all, they tapped millennial anxieties.

While no title predicted the specifics of 9/11, such a range of ludic dress rehearsals outlined at the very least an artistic milieu of the 1990s rich in disaster iconography and a cultural trajectory that seemed to be readying the nation for far-reaching disaster. Like Steve Jackson's *Illuminati* (1995) game cards—with one card depicting planes hitting the World Trade Center—a strong sense of prerecognition, coincidence, and prediction surrounded games such as *Deus Ex*.[42] A term associated with conspiracy theorists, "predictive programming," denotes how people can be acclimatized to future events and social changes by small references deliberately hidden in popular media. While in no sense an act of conspiracy, video games had nonetheless begun training people how to react to a terrorist attack, psychologically conditioning them for an impending disaster.

Together, *Parasite Eve*, *Urban Strike*, and *Deus Ex* fit the broader descriptor of premonition games. They explored scenarios much like September 11 and offered "dress rehearsals" of terrorist attack. They foresaw national disaster. In 1988 box art, *Microsoft Flight Simulator* depicted a commercial plane flying out from the Twin Towers, while artwork on Gottlieb's pinball machine *Rescue 911* (1994), based on the CBS television show, featured scenes of ambulances responding to a New York-like city on fire.[43] Sometimes the proximity between "premonition" and "real" could be disturbingly uncanny. Throwaway images of destruction pre-9/11 suddenly took on significance after the attack, either as newly offensive material or as strangely prescient and informative. Evidence of "image repetition" of the Twin Towers falling thus existed years before repeat footage of the real planes hitting in 2001 circulated. The attacks of 9/11 seemed to exist as coded fragments in games before they became a reality.

Games prepared audiences for catastrophes like 9/11 in a number of subtle ways. They provided guidelines of perception. Cultural images of disaster preceded the reality of 9/11 in a programmatic sequence. The power of the (gamic) image hit first; it established the scene and what to expect before the reality followed, increasing the impact. For Baudrillard, "the real is superadded to the image like a bonus of terror, like an additional frisson."[44] The 9/11 terrorist attacks built on well-established iconography. Films, literature, and games provided the conceptual guidebooks for the disaster, as they had done for other historic events. Released March 14, 1979, the atomic thriller *The China Syndrome* starring Michael Douglas, Jack Lemmon, and Jane Fonda, fictionalized a nuclear power plant accident on the California coast. On March 28, 1979, a partial reactor meltdown occurred at Three Mile Island nuclear plant in Pennsylvania. Many citizens made sense of the Three Mile Island accident through the cinematic lens, the movie explaining scientific jargon, nuclear dangers, and emergency protocol to its audience in a far superior way to that of bumbling, confused officials on site.[45] Images in popular culture, especially fictional depictions of disaster, served as a valuable conduit to understanding and comprehending.

Premonition images also made disasters like 9/11 familiar, normal, and psychologically survivable. Repeat images of disaster in both game and film made real disaster yet another image. Like an anesthetic injection, they prepared the patient for pain by numbing them to it.

Compared with other entertainment media, pre-9/11 games offered a distinctive sense of being present. They offered a sense of interaction. Games allowed players to explore dangerous scenarios first-hand. People became participants, not voyeurs, in image/events. Specifically, games offered the attractive proposition of not just being spectators for "the end," but having a role in it, whether facilitating Armageddon or preventing it from happening. A key activity of gaming, tackling destruction was crucial to gamer experience dating back to the 1970s. In *Space Invaders* and *Galaxians*, players fended of brutal alien attacks. In the Nintendo 64 title *Blast Corps* (1997), players operated a range of vehicles to clear the way for a nuclear warhead traveling on a carrier on an automated course. *Blast Corps* encouraged players to destroy buildings at will in order to prevent a much larger explosion.[46] Video games gave audiences a means to play with their doomsday imaginations.

Games also connected the fin de siècle moment and the gateway of New York. Max Page notes, "Our popular culture has been a dress rehearsal for the city's destruction for decades: in books, at the movies, in computer games." Games such as *Deus Ex* cast New York as a historic symbol of disaster and underlined its vulnerability at the close of the millennium. Software companies gamified cultural concern over the fate of Manhattan skyscrapers. Falling skyscrapers were a common fear in the public mindset. As early as 1964, critics of plans for the World Trade Center pointed to the potential for airplane damage, with reassurances by structural experts that any accident would remain "local" and "occupants outside the immediate area of impact would not be endangered."[47] Page notes that the Committee for a Reasonable World Trade Center argued for a reduced height out of fear of an air crash. For Page, skyscrapers were "eerily present" in disaster fiction leading up to 9/11 and popular targets for "monsters, bombs and comic book heroes." The World Trade Center was targeted for destruction pre-destruction, and seemed part of a broader New York end-game.[48]

Finally, games interpreted disaster as fundamentally about America's fall. The fictional World Trade Center stood and fell as a part of a larger endgame. A vertical barometer of global threat, New York skyscrapers faced submersion in *Deep Impact* (1998) and a tsunami in *Independence Day* (1996). They similarly fell or simply disappeared in a range of video games, from *New York Blitz* to *Deus Ex*. Games, film, and literature alike fictionalized the end of America through the lens of New York. American image makers manufactured and sold disaster. As Slavoj Žižek contends, "America got what it fantasised about," with fears of the destruction of sacred places coming to fruition on September 11.[49] Baudrillard highlighted "the fact that we have dreamt of this event, that everyone without exception has dreamt of it—because no one can avoid dreaming of the destruction of any power that has become hegemonic to this degree." Terrorism seemed the murky shadow or inevitable side effect of the mammoth rise and imagined fall of America. Baudrillard raised the issue of collective guilt for imagining the attack ahead of time, of "deep seated complicity," in that "we can say that they did it, but we wished for it." As Baudrillard contends, "disaster movies bear witness to this fantasy," but games imagined it too.[50] Like an updated Greek or Roman tragedy, cultural stories charted the rise and fall of empires and skyscrapers alike. The loss of the Twin Towers in games provided a convenient symbol of an altogether greater loss.

When the actual World Trade Center collapsed on the morning of September 11, the immediate reaction was to delete all existing cultural imagery of it. As 9/11 unfolded, as the concrete and steel fell, the standing towers were exorcised from the popular zeitgeist. Like a defunct computer code, the two towers were deleted from stored memory. Destruction read deletion. As Thomas Stubblefield elaborated, "absence, erasure, and invisibility dominated the frame."[51] Any representation of the standing towers became instantly controversial. Television channels deleted World Trade Center scenes when showing movies such as *Home Alone 2: Lost in New York* (1992). The Hollywood movie industry responded to images of 9/11 with strident acts of self-censorship. Prior to release, a final scene in *Men in Black II* (2002) shot in front of the World Trade Center was replaced with a less controversial denouement at the Statue of Liberty. One architectural symbol was exchanged with another. Editors deleted a scene revolving around the Twin Towers in Sam Raimi's *Spider-Man* (2002) picture. Fearing that their movies might reignite trauma, film studios deemed any representation of the World Trade Center as potentially offensive to audiences. The attacks of 9/11 spelled the elimination of Twin Towers from the contemporary film world. It seemed almost as though the towers had fallen onto the film-editing floor.

The immediate reaction of the game industry mirrored Hollywood in its visual disavowal. Wary of courting public criticism, design companies judged 9/11 too traumatic to pass as entertainment. No 9/11 imagery, Twin Towers or otherwise, was allowed into mainstream video game content. Programmers carried out a cleansing action similar to that of film editors. The attacks of 9/11 became a non-event, a structure with no presence or place in computer code. Activision deleted the sky-scrapers from *Spider-Man 2: Enter Electro* (2001) for the Sony Play-Station. Three days after the terrorist attack, Microsoft announced deletion of the World Trade Center from its flight simulation game, and the 2002 edition had all building damage options removed. Sega cancelled its flying game *Propeller Arena*. Sony delayed the release of spy game *Syphon Filter 3* (2001), with Ami Blaire, director of product marketing, casting the title as "too sensitive to introduce during this time of tragedy."[52] Changes to Rockstar's New York–based *Grand Theft Auto III* (2001) included replacing New York Police Department vehicles with generic police cars, removing a terrorist-related mission, and redirecting a plane away from skyscrapers in one scene.[53] Electronic Arts repackaged *Command & Conquer: Red Alert 2* because it depicted an attack on New York City. The company removed box art that showed a skyline (including the Twin Towers) on fire and a plane heading toward a burning World Trade Center.

Originally called *MGS III* because of a plot revolving around Manhattan's three highest skyscrapers, *Metal Gear Solid 2: Sons of Liberty* (2001) also had sections rewritten after 9/11. Drawing inspiration from Paul Auster's *City of Glass*, James Ellroy's *L.A. Confidential*, and real-world terrorist attacks, *Metal Gear Solid 2* had been in development since early 1999. The game presented a sophisticated narrative of world terror. Targeted at the American market, it presented New York City as under threat from multiple factors and antagonists: the collapse of the stock exchange, escalation in nuclear weaponry, enveloping pollution, a terrorist threat, and a governmental conspiracy. Consciously scripted so that "the player is unable to tell fact from fiction," *Metal Gear Solid 2* blurred the line between fantasy and authenticity. In response to 9/11, the game's programmers deleted a traumatic scene in which weaponry destroyed the Statue of Liberty and Manhattan's Financial District. An action section involving the Twin Towers was also abandoned. The Twin Towers disappeared from gamic New York.[54]

Game companies defended the erasure of the Twin Towers by the logic of architectural accuracy and the emergence of a new New York City. Companies avoided issues of national trauma by excluding any reference to the attacks. They sidestepped emotional impact. Companies showed respect to the "hallowed ground" of 9/11 by not intruding on it or codifying it.[55] The attacks of 9/11 were not to be played with;

they remained immutable and sacrosanct. The wholesale negation reflected a reticence, even fear, over manipulating disaster imagery and of appending concepts of play to mass death. The real Twin Towers had been destroyed and violently erased. The moniker Ground Zero for the space where the Twin Towers had once stood, a hijacking of a nuclear term, became the only apposite description for the sense of total annihilation and obliteration. A sense of nothing left at Ground Zero dominated. Nothingness and invisibility replaced a built skyline. Like other media, video games respected the emptiness. For the gaming community, 9/11 thus invited a rare moment of sensitivity, pause, and respect. Such reflection intimated an entertainment industry finally maturing, leaving behind its reputation for graphic as well as graphical violence, or as Karen Engle describes it, the ability of games to be "excessive, indecent, or unwarranted" with their images.[56]

However, the wholesale deletion of the World Trade Center from the entertainment world offended as well as calmed. Referencing the Ben Stiller comedy *Zoolander* (2001), Todd McCarthy at film trade magazine *Variety* railed, "Deletion of the towers from the picture is infinitely more disruptive, not to mention insulting, than leaving them in."[57] Critics pointed to the abrupt and unnecessary erasure of an American symbol, as if the culture industry had collectively and autocratically decided to delete the White House or Yosemite National Park. The deletion allowed popular culture to continue as if nothing had happened. Films and games thus continued to offer a fantastical escape, with the magic circle surviving the terrorist attack largely intact.

Games (supposedly) made no contribution to the growing imagery surrounding 9/11. That the industry offered no clear perspective on events left a curious and lamentable void. As Don DeLillo wrote, in the absence of alternate stories, "the world narrative belongs to terrorists."[58] The game industry seemed impotent in challenging such a narrative or conceptualizing 9/11 in virtual form. It seemed as though the fantasy/entertainment realm of digital gaming could not connect with the new reality dawning.

The act of deletion also raised the issue of what to put in the negated space. Tackling largely film, Stubblefield argued that the lasting "empty image" of 9/11 remained crucial and that "absence functions not as negativity but as a particular mode of presence which shapes experience and official histories in often dramatic fashion."[59]

Only a few titles challenged the mass cultural censure and dared to put the Twin Towers on screen. In 2003, independent production company Kinetic produced a modified patch for the *Unreal* game engine that transformed the PC title into *9/11 Survivor* (2003). Kinetic adapted a standard first-person shooter, converting space monsters into terrorists thanks to a list of 9/11 codes that could be added to the program. In the French-designed flash game *New York Defender* (2002), players tackled a terrorist attack on the Twin Towers by shooting down incoming planes. Gamers stared at the distant New York skyline through crosshairs, with the World Trade Center at the center point, and attempted to take down a range of planes coming in at variant angles and velocities from the edge of the screen. The game featured the pre-game warning "may offend some." In its old-style graphics and antiquated gameplay, the title vaguely resembled *Missile Command*. Players simply aimed and fired. Instructions noted how "we all remember the mournful day of September 11" and lamented the lack of opportunity for citizens to stop the actual attack. *New York Defender* claimed to offer the player the chance to "return back in time" and "create your own alternative history." Rather than feeling like a helpless victim, the player could instead "destroy the terrorists at least in the virtual world" and "feel yourself like a hero." The narrative promised psychological retribution, even rebirth. In brutal gamic simplicity, *New York Defender* captured some of the most common issues connected with 9/11: personal impotency, what could have been, and the desire to be a hero. The repetitive element of the game meanwhile captured a familiar sense of the fall of the towers portrayed in popular media imagery. The waves of terrorist planes never stopped, with the threat seemingly never-ending. Towers kept falling like they did on television screens. As in *Missile Command*, the game posed the issue of inevitable defeat. Nobody could "win"; players could only gain points for the number of terrorist planes shot down. The destruction of the towers was pre-programmed and fully automated. Like *Missile Command*, *New York Defender* highlighted the unwinnable nature of conflict and situated the fight against terrorism as fundamentally flawed.[60]

Around 2006, the embargo on the Twin Towers in film and literature expired. As with other conflicts, the period of "trauma time" ended. Although over a decade passed before the Vietnam War was critiqued by filmmakers, artists and writers, 9/11 took just five years. Don DeLillo's novel *Falling Man* (2007) and Oliver Stone's movie *World Trade Center* (2006) marked 9/11 as legitimate grounds for artistic inquiry. Architectural nostalgia for the Twin Towers grew, with a mass sense of needing to preserve the memory of the old New York skyline. The American public seemed to be moving through the phases of mourning: from the delusion of permanence, to the shock surrounding death, to the preservation of memory. Work began on what to put at Ground Zero as a memorial and to form a lasting replacement image.[61]

Explorations of 9/11 in the game world nonetheless continued to prove complicated. In a largely artistic piece, *Invaders!* (2008), Douglas Edric Stanley, an artist raised in Silicon Valley, superimposed the classic arcade game *Space Invaders* on a backdrop of the Twin Towers as an installation at the 2008 Leipzig Games Convention. Pixelated invaders made their way down a giant screen featuring an image of the towers. The invaders posed as space terrorists. Gallery visitors played the game like a Nintendo Wii title, by using hand gestures and waving their fingers to stop the onslaught. The game never stopped, with instead the individual energies of participants waning. Like *New York Defender*, *Invaders!* highlighted the War on Terror as flawed and served as a protest statement against US foreign policy. However, the art-game was a perplexing experience. Players failed to reconcile the seriousness of the visuals and message with the frivolous physicality of hand-clapping. In fact, they often laughed as the towers fell. *Invaders!* created not only ludic dissonance, a paradox of play (to win points but lose lives at the same time), but also a paradox of emotions.[62]

For the mainstream game industry, the question remained how to tackle 9/11. Intriguingly, companies still responded by avoiding any direct reference. The Twin Towers remained mostly absent from game worlds. Instead, the fallout of 9/11 emerged in the broader digital landscape. All sense of gamic New York City as a friendly tourist landscape disappeared. Instead, the sense of Manhattan as a city under attack grew exponentially. The narrative void left by the destruction of the Twin Towers gradually filled with conspiracy theories, apocalyptic battles, and a long-term digital war on terror. Games revisited themes of disaster and subterfuge first explored in *Deus Ex*. Programmers collectively recoded the city. New game franchises *Call of Duty*, *Resistance* (2006+), and *Crysis* (2007+) all at some point explored fictional attacks on New York City. In team-based shooter *Tom Clancy's The Division* (2016) by

Massive Entertainment, New Yorkers faced quarantine, smallpox, and "dollar flu," conjuring images of terrorist-distributed anthrax. Game designers used the shadow of 9/11 to explore greater fictional scenarios of America destroyed in *The Last of Us*, *Homefront*, and the *Fallout* series. Extending "ground zero" well beyond Manhattan, programmers explored a whole digital nation under attack. Game designers expanded the fallout of 9/11 past the blast zone, and into new fictive realms. The cultural processing of Ground Zero manifested in a growing range of apocalyptic game settings, with the ashes of terrorism converted into new digitized derelict landscapes. Mainstream game companies presented America as a permanently embattled landscape.[63]

Although games did not address 9/11 head on, the specter of the Twin Towers influenced gameplay. The fall of the two skyscrapers determined subtle shapes and forms in the new digital detritus. The remains of the World Trade Center could be seen in the rubble of an alien-defiled New York in Crytek's *Crysis 2* (2011). The player looked out onto a collapsed city made up of fallen buildings, broken concrete, and universal grayness. The overwhelming sense of emptiness and decay was reminiscent of an evacuated and frightened 9/11 New York. Visiting this new gamic Manhattan for some meant revisiting old trauma. As one *Gamespot* blogger wrote, "as a New Yorker and a gamer I find that New York is a great place to place stories in but it does remind me what happened during 9/11." Another gamer felt that software companies were out to "tap into cheap emotions."[64]

In a sense, the towers were still present thanks to so many conspicuous virtual ruins. Consciously or not, programmers crafted thinly veiled facsimiles of 9/11 destruction. The long shadow of the Twin Towers endured, extending out to mark New York as one mass grave. Every gamic presentation of New York City payed homage to the rubble on some level. Reflecting on the meaning of the fall of the World Trade Center, Jean Baudrillard suggested, "Their end in material space has borne them off into a definitive imaginary space."[65] Stubblefield connected the destruction of the World Trade Center with the fall of the statue of Saddam Hussein outside Baghdad and the rise of the spectacle of iconoclasm.[66] The attacks of 9/11 assured the mythic status of the Twin Towers as twentieth-century statues and urban wonders. *Architecture Week* once labelled the two skyscrapers "cathedrals of our age," and the spirit of the Twin Towers proved hard to extinguish.[67] Symbolic presence remained despite their physical absence. Ephemeral, hinted at, and sometimes on the edge of being revealed, the Twin Towers provided a structural haunting in game worlds. The continual implosion of digital buildings was arguably part of the same cycle of repeat imagery that first began on September 11 with the Twin Towers falling over and over again on television screens. Like a burn in photographic film, the Twin Towers kept coming through.

Simulation and the Falling Man

Mitchell called 9/11 a "traumatizing spectacle."[68] With the new popularity of camera phones and digital cameras, the explosion of the Twin Towers was an image/event caught on thousands of devices by thousands of spectators. Out covering a fashion shoot in Manhattan, photographer Richard Drew, like many, turned his camera to the towers. His attention was caught by people jumping from the north tower, in acts of desperation, faced with suffocation if they stayed inside. He shot several reels before falling debris forced him from the scene. The *New York Times* published one of Drew's pictures the following day on page 7. The image became known simply as "The Falling Man."[69]

The image of "The Falling Man" appeared, then disappeared. It was captured, then lost. After featuring widely in initial coverage, it went missing from articles, documentaries, dramas, and reports that followed for several years after. Like the Twin Towers, the jumper was deleted from popular culture: the visible was made invisible. People looked elsewhere for images of death and systems of grief and mourning.

The deletion reflected a variety of rationales, some obvious, some hidden. The Falling Man was above all a discomforting image and soon became recognized as a "traumatizing spectacle" too shocking to publish. Abiding by Jeffrey Melnick's "auto-censorship" theory, newspapers voluntarily withdrew the picture.[70] Early censorship of the image reflected the desire to show respect to the victims and their families. Around 200 people jumped, mostly from the north tower. "Seeing" a family member die represented the ultimate traumatic image. The Falling Man was inevitably someone's partner, son, or father.

The existence of the Falling Man also conflicted with the cinematic image of 9/11, as well as a movie-led understanding of the event. While movies and games prepared America for 9/11, none showed actual death.

The series of falling shots cut through the filmic 9/11 like acetate. They undercut a safe, voyeuristic film event by drawing attention to the painfully real. An outside lens created some psychological distance, while the Falling Man image provided an intimate focus on death. Death on 9/11, like a classic movie, was meant to be something that happened "off-screen."[71]

Editors deemed the image too powerful even for 9/11. Artistically, it connected the individual to the monumental: the falling of the planes, the towers, and then the people. It reflected the visceral reality of the death of thousands, of actual carnage, and provided a potent symbol of personal loss. It was outside the acceptable range of emergent 9/11 iconography: an exception, a human death, among images of building detritus. As *Time* magazine commented, "On a day of mass tragedy, Falling Man is one of the only widely seen pictures that shows someone dying."[72] Such intensely traumatic imagery sat outside the parameters of normative American visual culture. For Žižek, an absence of images of bodies at Manhattan showed the "derealisation" of horror at the World Trade Center: that "real horror" still lay abroad, in news coverage of distant famines and wars, and was not something that happened on US soil.[73] The Falling Man image contradicted rules over the seen and the unseen. The events of 9/11 represented something bigger than the individual, bigger than the Falling Man, so no one wanted to see him or to know him. The real symbols of the tragedy were much larger. They were metal and mechanical, not organic and bodily. The fall of the Twin Towers provided the quintessential image of national loss.

The Falling Man image similarly challenged popular attitudes toward death. The photographic image portrayed both individual and mass suicide. With suicide stigmatized in Western culture and widely associated with failure, the Falling Man became an image of national embarrassment. Its existence challenged the singular narrative of "heroes" that was fast emerging around 9/11. Instead, it depicted Americans as vulnerable, even desperate. At odds with the images of survivors and heroes, the Falling Man moved in the opposite direction to the tide. Moreover, suicide provided an unwelcome connection to the bombers themselves, universally cast as cowards in the West. Discomfort over the image lingered for some time. Reporter Tom Junod searched for the family of the Falling Man to no avail.[74] He was likely a worker at the Windows of the World restaurant near the top of the tower, but embarrassed families vehemently denied any connection. A suicide case could not be a proud family member. "Falling Man" was nobody's partner, son, or father.

On many levels, the Falling Man simply did not fit with the narrative around 9/11. In its unworldly, abstract composition, Richard Drew's photograph emphasized a figure isolated from everything else. The Falling Man appeared untethered, not even connected to the towers, let alone a terrorist attack or world event. Most importantly, the Falling Man failed to meet the cut when it came to creating and patrolling the fast-emerging iconography of 9/11. As a set of images solidified around the attack, the Falling Man fell to the ground, discarded as a pariah. The Falling Man became victim again, not of terrorism, but of the desire to control imagery.

The reality was that the Falling Man fit in with a long history of photography, and connected with other historic events in New York. The photographic style of the multiple, repeat images of movement taken by Drew resembled Eadweard Muybridge's early photographs of horses. While itself a static image, the Falling Man, as part of a series, was imbued with motion and speed. In more local terms, Stubblefield placed the Falling Man within a framework of New York suicides from the past, both real and fabricated.[75] During the Wall Street Crash of 1929, stories circulated of mass suicides, of destitute bankers jumping off Manhattan skyscrapers; these reports were largely erroneous, although newspapers suggested otherwise. Jumping to one's death also became a filmic and a literary device, with D. W. Griffith's *The Birth of a Nation* (1915) showing a white girl commit suicide rather than be captured by an African-American slave. More broadly, the fall of man—both literal and figurative—represented a dramatic fall from grace. For Baudrillard, the Twin Towers themselves were caught on film committing an act of suicide by their fall.[76]

The Falling Man similarly connected to imagery in video games. Statistically, falling is rarely used as a means of suicide. By contrast, in gaming circles, falling is a commonplace means of dying. While the Falling Man represented an action rarely seen in reality, the act of jumping to one's death was typical of several decades of play. Since the 1970s, gamers had jumped to their digital deaths in titles such as *Pitfall!* (1982) and *Spider-Man* (1982), watching their pixelated, third-person avatars plunge down the screen and disappear, often in slow motion. In its stop-frame, almost rewindable state, Falling Man seemed little different from these depictions of gamic death. The Falling Man photographs resembled a series of screenshots, with the man resembling a pixelated player. Generic looking and slightly blurry, with no clear identity, he seemed indistinguishable from a game character. Games coded the act of falling uncontrollably as failure and death. In code didacticism, to fall meant to fail, and jumping a long way led to death. Platform titles such as Sega's *Sonic the Hedgehog* (1991) and Nintendo's *Super Mario* (1985+) highlighted the danger of falling as a chief obstacle to progress. However, platform games also rewarded the player with multiple lives. Through the continual return of the avatar, software companies situated death by falling as a hindrance to play, but not always the end of it (represented by a "Game Over" screen). Often a painless experience, losing a life happened frequently and without deeper meaning.[77]

In the 2000s, increasingly realistic graphics closed the aesthetic distance between the Falling Man and the falling avatar. A brief, inconsequential pause before restarting a game level, gamic death suddenly became closer to the real experience.[78] Something hidden became eminently explorable. In first-person shooters such as *Destiny* (2014), falling induced a sense of vertigo and loss. In EA DICE's *Mirror's Edge* (2008), set in a dystopian future of totalitarian governance, the player assumed control of Faith, a messenger for covert information, who used her parkour skills to navigate city obstacles, including skyscrapers. If the player missed just one key jump, Faith fell instantly to her death. For the player, the first-person experience of the fall invited feelings of panic, with Faith's vision blurring on screen, accompanied by the sound of wind, then nothingness. EA DICE provided no "witnessing" of Faith's body on the floor after falling. Turning the camera away at the last moment encouraged the player to imagine the death (mental imaging continuing where the graphics left off). *Mirror's Edge* encouraged actual sensations of falling, inviting the player to feel fear and experience first-hand sensations of vertigo. The game effectively simulated a "falling woman." Meanwhile, in the PlayStation VR title *The Walk* (2015) gamers faced the virtual experience of walking a tight rope between the Twin Towers as they existed in 1974.[79]

Being the Falling Man

Attempts to return the Falling Man to the mainstream imagery of 9/11 proved highly divisive and mostly failed. The Falling Man instead entered the world of shock art. In September 2002, artist Eric Fishl displayed his sculpture "Tumbling Woman" in the concourse of the Rockefeller Center in New York. Complaints of insensitivity, poor taste, and "too soon" led it to be quickly withdrawn.[80] "Falling," a "window project" by Sharon Paz, proved similarly contentious with its silhouettes of falling humans pasted onto the external facade of the Jamaica Center for Arts in Manhattan, also in September 2002.[81] Published six years after the attack, Don DeLillo offered perhaps the first acceptable version of the Falling Man in his short novel *Falling Man* (2007). In DeLillo's novel, the falling man served as a side character in a greater exploration of trauma, focused on a performance artist in New York, in business garb, who suspends himself from a wire. A failing safety harness meant the performance always had problems. The character also suffered from an addiction to repetition.

Treating the Falling Man in a similar way to the Twin Towers, the commercial video game industry held down the delete button. Software companies avoided such a serious subject, and by their uniform silence, helped patrol the acceptable imagery of 9/11. While a few generic smartphone games asked players to maneuver falling men across levels of variant difficulty, none were linked directly to the terrorist attack. Only one title consciously engaged with suicide, jumping, and the 9/11 experience. Part of a French college project overseen by creative director Anthony Kraft, the virtual reality game *08.46* (2010), or "9/11 Simulator," for the Oculus Rift headset tested the boundaries of public acceptability as much as the limits of new technology. Set on floor 101 of the north tower, the demonstration game cast the player in the role of an office worker at their computer during 9/11. While office manager Audrey talks on the phone, a sudden noise and building sway indicates a hijacked plane hitting the building. A fire marshal appears, and the player attempts to escape, navigating a series of locked doors and increasingly frightened colleagues. On entering a final office room, a man is seen dialing 911 to no avail. Amid the growing smoke, he breaks a window and jumps. The player watches as the falling man descends to his death. The game camera forces the player to engage with the event, unable to look away. Audrey duly suffocates. The player then faces a choice: either stay inside the room and suffocate like Audrey, or jump out the window like the man. The opportunity to jump from the north tower, to "become" the falling man, represented a rare moment of experiential simulation. Either option led to the Game Over credits.[82]

Like gazing deeply at a photograph, *08.46* offered the chance to "enter" 9/11 as an image/event. The power of the game derived from its use of virtual reality technology. The VR headset enabled people to visit the Twin Towers for themselves. By using the latest visuals and technology, *08.46* brought 9/11 once again into the home, but as a fully interactive and immersive experience. The title made 9/11 "real" for the individual player. Players noticed the subtle sounds of the office, the quiet immediately following the explosion, the darkness of corridors with only emergency lighting, the smoke engulfing the building, and the sense of panic felt by coworkers. The title highlighted the multiple sensations of that day. It enabled a richer, more disturbing experience than the usual, carefully framed visual footage.

The virtual reality experience touched on something palpable, non-technological, and primeval. It ignited emotion and memory. The game likely reopened the door to trauma for some. Through its intense, dramatic portrayal, the game captured Don DeLillo's "primal terror" of the 9/11 moment.[83] It provided a digital haunting and a sense of 9/11's ghosts. It served as a reactivation code for collective memory. By enabling everyone and anyone to "be there," *08.46* served as a powerful reminder of 9/11 as a mass experiential event. The events of 9/11 were a major event "seen" by all, but mostly through mediated newspaper images and television screens. Virtual reality transformed the experience of watching the disaster into being *in* the disaster. Casting the player as a victim of 9/11, *08.46* functioned as a digital exercise in imagined loss, both of self and others. Its brevity highlighted the shock of that morning and the unplanned nature of life. It offered the potential to explore death and mourning, as well as the mystery of the Falling Man. On some levels, the game offered 9/11 as a code never quite finished and a story worth revisiting.

The title was highly controversial. Denise Matuza, whose husband died in the north tower, exclaimed, "Why would anyone do this? It's disgusting. It hurts." One New York newspaper simply labelled it "disgusting."[84] While some applauded the realism of the virtual reality and the educational benefits offered, as well as noting the multitude of games based around other difficult topics (world wars, Pearl Harbor, etc.), others found the game "disrespectful" and asked, "What next[?] VR of a nazi [sic] camp?"[85] Appropriating any element of 9/11 for VR entertainment was likely to provoke controversy, but that programmers focused on the experience of the Falling Man and other north tower workers made it worse. The game threatened the established reading of the 9/11 image/event by re-opening it as virtual experience. The events of 9/11 were something to memorialize and sanctify, not for memory recoding or a game. Criticism also reflected a broader attitude toward video games and their limits. While literature and film had the potential to tackle trauma with some sensitivity, most people felt that games cheapened and essentialized important events. Attempts to tackle serious topics had failed in the past, with *Super Columbine Massacre RPG!* (2005), covering a US school shooting, widely condemned. That no one title tackled 9/11 effectively provided the requisite proof of the inferior status of video games. For most software designers, too, 9/11 was inappropriate content. By its associations with trauma, barbarity, and terrorism, 9/11 sat outside conventional notions of play spaces as realms of escape and mass relief. Most magic circles provided fantasies of individual heroism, not simulations of national tragedy. Traditions of play and the moral, contextual, and coded boundaries of game worlds cast 9/11 as too extreme. In the case of 9/11, games designers and critics alike defined video games as "entertainment only." Although, as David Altheide contends, popular culture serves an important purpose in navigating the meanings of 9/11 and in tackling ideas about grief, memorialization, and celebration, games were kept on the periphery of the discussion.[86]

The game *08.46* also presented a delusion and arguably a cognitive scam. Taking off the headset, the player instantly recognized the fiction at hand. Virtual reality technology promised authenticity, but *08.46* delivered only a fleeting, impressionistic view of proceedings. While players could see the havoc around them, they could not smell the smoke or feel the heat. As with a theme park ride, the game provided a recognizable simulation of fear, not to be confused with fear itself. Thus, the game highlighted the limits of virtual reality, with the player easily comprehending the game as simulation by its preprogrammed nature as well as its blocky graphics. The events of 9/11 remained impossible to fully replicate, assuring their status as part of a greater narrative of American exceptionalism. The game also revealed a broader discomfort with virtuality and simulation. Were some events best left alone in the virtual world? Guy Debord wrote in the late 1960s that "all that was once directly lived has now become mere representation."[87] Arguably, images of 9/11 were already mediated enough, without adding virtual reality programs and gameplay to the mix. To gamify 9/11 seemed a step too far. Like with theme parks, people wanted their video games to offer "fantasy realism," not perfect replication and the uncanny. Most gamers wanted to play in fantasy America, not real America.

Virtual reality games about 9/11 threatened the established relationship to the events of that day. In photography and film, a distance remained between viewer and image/event. Drawing on Susan Sontag's sense of the photograph as a "non-intervention" and a distance maker, Stubblefield noted how people at the scene on 9/11 employed the camera as a "defense mechanism which safely removes the subject from a scene that is too great."[88] Photography slowed down the action, put a (albeit minor) wall between person and place, and transformed the experiencer into the voyeur. Even at Ground Zero, the camera offered some protection from the full impact of 9/11 by offering locals the role of documenters or observers. The screen thus provided a protective medium. It protected against feelings of being the Falling Man or directly experiencing the horror. However, virtual reality games cast people as participants and interveners, not as observers. The classic magic circle of games enveloped players in an immersive experience. Rather than keeping events at a safe distance, virtual reality brought 9/11 into a new, fundamentally closer focus.

Games thus posed a danger to the conventional image of 9/11. With its fixed iconography, 9/11 existed as a carefully maintained and largely didactic diorama, one that told a story of national loss and national survival. By providing opportunities for revisiting, reshaping, and retelling the events of that day, games challenged any notion of singularity. Through their multiple plot threads and player freedoms, they threatened the accepted narrative of 9/11. As the Falling Man was cast out of the pantheon of acceptable 9/11 imagery, so too did game worlds such as *08.46* exist outside the "disaster circle."

Perhaps games needed to explore other sides of September 11. In Native American author Sherman Alexie's short story "Can I Get a Witness?" (2003), a fictional software designer, when asked by a Spokane Indian woman about his experience of September 11, admits that he had been working that day on a first-person shooter based around being a terrorist, with the World Trade Center as one of the levels. The woman asks, "How could you live with yourself?" He replies, "We redesigned the game after the eleventh. Now you play a cop who hunts terrorists." The company also deleted the World Trade Center level. The Spokane Indian initially responds, "That's blood money," but on reflection ponders, "It's tough to be open-minded about this stuff. But you've got to be. You can't let any event have one meaning, right? Your games don't have one meaning, do they?" The designer replies no. They conclude the conversation: "All right, maybe September eleventh means things nobody has thought of yet." The multiple paths of video games indicate that multiple understandings remain out there.

Firefighters, Flags, and Patriotism

Just after 5 pm on September 11, Thomas E. Franklin, working for the *Bergen County Record*, photographed three New York firemen hoisting a flag above the debris at the World Trade Center. The newspaper published the image on its front page the following day, titled simply "Firefighters Raising Flag."[89] Several other newspapers picked up the image, and it ran internationally. The picture was renamed "Raising the Flag at Ground Zero." Franklin's picture thus helped configure the visual culture of 9/11.

"Raising the Flag at Ground Zero" depicted hope in hard times. Diametrically opposed to the Falling Man image, with its downward spiral and allusion to hopelessness, the fireman image looked upward, toward the heavens and toward survival. Amid the dominant thematic imagery of falling—the towers, the people, and the planes—here was an image that pointed the other way. It provided a new tower of patriotic fervor. It suggested foundations that even terrorists could not destroy. The image was something to be shown, not hidden from view. It fit the controlled image of American unity, survival, and faith desired by the media, the public, and politicians. The raising of the flag represented the ultimate act of defiance. It offered hope and heroism in spades, and sidestepped issues of death and mourning. Franklin's picture resembled an older and equally potent photograph. In World War II, American photographer Joe Rosenthal produced a picture of six US marines holding the Stars and Stripes aloft Mount Suribachi during conflict. The image became known as "Raising the Flag at Iwo Jima." Visual similarities helped cast the 9/11 image as fundamentally about America at war again. The metaphoric power and nationalistic imagery of "Raising the Flag at Iwo Jima" transferred to Franklin's image. Franklin's photograph thus featured welcome pictorial, historic, and triumphant qualities. It carried a sense of continuity and permanence at a time of challenge.[90]

Mass media and commerce transformed Franklin's photograph into a mythic image. The picture of the New York firefighters was subtly reprogrammed with new subroutines, always with an eye on perfecting the heroic image. Sales of souvenirs on New York streets transformed the picture into a commodity, a memorial product to sell and an example of patriotism for purchase. On March 11, 2002, the White House used the image for one of the "Heroes 2001" US Postal Service ceremonial stamps. The stamp focused on the flag and brightened its colors. The New York Fire Department (FDNY) planned a sculpture for its Brooklyn headquarters, replacing the white crew from the picture with a more ethnically diverse fire team. A forty-foot bronze statue "To Lift a Nation" was raised at the National Fallen Firefighters Memorial Park in Maryland.[91] The FDNY Memorial Wall across from Ground Zero meanwhile featured an expansive bronzed mural depicting a scene of mass sacrifice for visitors to gaze at.

The 9/11 hero could be seen in all manner of media productions. Taking care to present September 11 in an uncontroversial light, Hollywood directors focused on the activities of emergency response teams. Oliver Stone's *World Trade Center* (2006) followed the work of Port Authority police officers and featured actual FDNY firemen onscreen, while the documentary *9/11* (2002) focused on firefighters and other lifesavers.[92] The celluloid firefighter helped frame 9/11 as something positive and heroic. Their deployment represented a broader fight to save America.

Video games similarly explored the firefighter and flag imagery of 9/11. The few firefighting titles prior to 9/11 focused purely on the technical aspects of the job. In the Atari arcade game *Fire Truck* (1978), two players sat in a cabinet that resembled a vintage fire truck and together "drove" it through city streets to reach a blaze. *Fire Truck* provided an early example of cooperative play. The cabinet included actual bells and horns, with no ludic function other than furthering the sense of immersion. In the Nintendo handheld Game & Watch title *Fire* (1980), the player controlled two firefighters maneuvering a large stretcher to catch babies jumping out of a burning building. Based on the 1977 blockbuster disaster movie starring Paul Newman and Steve McQueen, the Atari VCS home console adaptation of *Towering Inferno*, published in 1982, set the player the task of liberating trapped people from a doomed skyscraper. The instructions gave little detail about the cause of the incident, merely stating that a "downtown skyscraper has burst into flames."[93] Rather than a time meter, the game employed an innovative "people meter," with slow to act players losing both game time and desperate people caught in the flames. Reviewers applauded the action title for "achiev[ing] a respectable level of excitement without having a shot fired in anger."[94]

After 9/11, the gamic firefighter took on a more expansive, celebratory, and culturally significant role. Games designers interpreted the firefighter figure as the ultimate hero. Software recoded the modus operandi of the rescue service, putting it to work against all kinds of dangers. The officially licensed *FDNY American Hero Firefighter* (2002), *My Hero Fireman* (2008), *Real Heroes Firefighter* (2009), and *Firefighter 3D: The City Hero* (2016) all portrayed fire work as intrepid and courageous. Games explored the sacrifice and salvation of being a firefighter. In *Real Heroes*, one stage involved rescuing people on the 37th floor of a flaming and exploding skyscraper in a state of imminent collapse.[95] With clear allusions to 9/11, scenes included those of petrified officer workers in the dark (crying "Help. I can barely breathe.") relating their intense fears of death (one asks, "Tell my girlfriend I love her.") and trapped rescuers struggling to escape the flames, thanks to their "reckless heroics." The game captured a sense of media spectacle surrounding the blaze, with reporters trapped inside comforting each other with lines such as "think of the ratings." A helicopter ride offered a welcome "last train out of Dodge" for some. Throughout the disaster, the player, as fireman, represented the true hope of all. As another game, *Firefighter 3D*, attested, the service manifested the desire to "be the hero."[96]

Firefighters also became characters in more expansive and wide-ranging stories. In *Resistance: Burning Skies* (2012), the player assumed control of a firefighter with very little firefighting duties, but with the larger responsibility of saving his city. Set in 1951, the game explored a fictional attack on New York City by Chimera, terroristic aliens at first believed to herald from Russian experimentation, thus combining both

Cold War and War on Terror periodization. The story began with flames engulfing State Siland power station. Tom Riley, a New York firefighter, presented as an ordinary Joe, rescues a trapped colleague. Riley goes on to emerge as a national hero. He traverses the snowy ruins of the city, fighting Chimera at Times Square, in Central Park, in the Chrysler building, and at the Statue of Liberty. The 1951 dateline assured the absence of the yet-to-be-built World Trade Center.[97]

Post-9/11 games also tackled the hoisting of the Stars and Stripes. Initially, companies were reticent to feature the US flag prominently in games. In *MGS 2: Sons of Liberty*, Konami deleted an ending sequence in which flags fell spectacularly across the Manhattan downtown. Westwood Pacific deleted the flag from *Command & Conquer: Red Alert 2* cover art and replaced it with an atomic mushroom cloud.[98] No games recreated the moment captured in "Raising the Flag at Ground Zero."

Gradually, the flag embargo passed. Games joined other media in displays of patriotic endeavor. Released in 2009, *Freedom Fighters* for the PlayStation 3 highlighted the sense of "everyday heroes" and "patriotic service" caught up in the contemporary fireman image. The player assumed the role of Christopher Stone, a downbeat plumber and New York everyman, who gradually becomes part of a guerrilla faction. The first scene depicted a takeover of Manhattan by a tightly organized terror cell. In a series of images clearly inspired by 9/11, helicopters shot at a New York skyscraper, people fled, and smoke engulfed the city. Paying homage to the film *Red Dawn* (1984), in which ordinary Americans fight a Soviet invasion, everyday Americans emerged as surprise obstacle to the occupying force. Christopher's transformation from plumber to the "new profession now . . . freedom fighter" attracted Isabella, a beautiful and mysterious woman, and also a spy. "Freedom" in the game was heavily associated with the rise and fall of the American flag. The game map highlighted "home" with an image of the Stars and Stripes. The first mission involved the rescue of Isabella from the Brooklyn Police Station. The player then systematically retook districts of New York, literally hoisting flags above key buildings to secure them. The Star and Stripes restored actual and symbolic power to Manhattan. As on 9/11, the act of raising the flag served as riposte to the terrorist threat. As with other games, the title combined the dual threats of Soviet invasion and Islamic terrorist attack, fusing together the Cold War and War on Terror eras.[99]

Saving New York became a common gamic act following 9/11. Players assumed the roles of city saviors with a variety of vocational backgrounds, from DEA agents to Christian evangelists. While the appearance of avatars varied, post-9/11 titles nonetheless presented the player with a uniform heroic role, consistent with the archetype of the 9/11 fireman. By inviting the player to become America's champion, software companies provided the potential to subtly reconfigure the 9/11 experience and even recode trauma. As one blogger commented, games worked by the logic that "X took us down to our knees. Now YOU can be the hero to bring us back up."[100] In post-9/11 game worlds, an initial experience of overwhelming destruction led to the opportunity for a similarly mammoth resurrection. The phoenix would arise from the ashes. Looking at movies such as *The Day After Tomorrow* (2004), Stubblefield argued, "post 9/11 cinema prompts the viewer to lead humanity out of the apocalypse via an onscreen idealized self."[101] Post-9/11 gaming similarly prompted players to lead the way.

Marc Ouellette and Jason Thompson note how gamers thus became "citizen soldiers" responsible for both protecting the city and broader American freedoms in the "war theatre" of the virtual Big Apple. For Ouellette and Thompson, "When it appears in post-9/11 games, then, New York is no longer just a background, but a highly charged symbol of the country, specifically of the country attacked."[102] Following 9/11, the depiction of New York became that of a city under attack. Gamers indulged in a range of fantasy scenarios all inspired by 9/11. In *Call of Duty: Modern Warfare 3* (2011), the player assisted a Delta Force team repelling Russian terrorists at the Wall Street Stock Exchange. The team shouted defiantly "We Can't Lose New York," evoking memories of the firefighters at the North Tower. While some titles presented Manhattan as an indestructible landscape, others showed it in perpetual ruin, on fire, and in decay. The player's fight typically seemed multi-dimensional: for themselves, for New York, and also for America. The three roles proved synonymous and little differentiated. At whatever cost, all three needed saving. How to achieve that goal typically revolved around one core activity: fighting. Saving New York very quickly turned into fighting for New York. Seamlessly the player moved from "under attack" to simply "at war."

5

Fighting the Virtual War on Terror

On September 11, 2001, the Manhattan skyline became the front of
the new war.

PAUL VIRILIO, Ground Zero, 82

T HE ATTACKS ON 9/11 served as the ground zero of a new global
war. For some, including the Bush administration, the September
attacks justified a comprehensive War on Terror, entailing the invasions
of Afghanistan (2001) and Iraq (2003), the establishment of a detention
camp at Guantanamo Bay (2002), as well as leading to reprisals by ISIS
around the world. For critics of the Bush administration, the War on
Terror represented a distasteful exploitation of national tragedy. As
Karen Engle stated, "the con of the towers consists of their disappearance being made to justify invasion, murder, and torture."[1]

The absence of 9/11 in mainstream gaming represented a conceptual
void. Like the crater at Manhattan's Ground Zero, 9/11 left a gap in
game code. How could the industry deal with the most important moment of the twenty-first century? While a number of titles had presaged
9/11 by imagining a terrorist attack on home soil, software companies
struggled to respond to the actual event. Shocked by the chaos at Ground
Zero, the industry momentarily paused. It then largely committed to a
new direction. Programmers directed the lens away from catastrophe,

so that the real and the horrific, the fall of the Twin Towers and the Falling Man, mostly disappeared from games. They redirected attention to something more palatable. New commercial games focused on variations of the War on Terror, offering distant crusades against generic enemies and the fantasy of winning. They codified 9/11 as a route toward national renewal and military victory.[2]

Fighting for America

Used to gamifying historic events as scenarios or levels to win, the industry initially tackled 9/11 by mostly skipping the level altogether. Both Hollywood and the US game industry found it easier and less risky to produce stories about the conflict that followed the attacks, rather than tackle events at Ground Zero. Military movies *The Sum of All Fears* (2002) and *Black Hawk Down* (2001) met with box office success, while first-person shooters *Call of Duty* (2003+) and new versions of Valve's *Counter-Strike* (2000+), a popular multiplayer shooter, provided a gamic counter-narrative to the sense of terrorists winning. Whereas Don DeLillo observed a "world narrative" taken over by terrorists on the morning of 9/11, military-based video games imagined otherwise and gradually coded a narrative of American heroes.[3] Themes of militarism, patriotism, unity, and heroic sacrifice abounded across a range of blockbuster action titles. The commercial game industry offered a welcome alternative narrative of contemporary history, in which players always won against terrorists. Games offered digital opportunities for mass retribution.[4]

Video games depicted the War on Terror in a wide range of conflict zones and virtual territories. Campaigns moved beyond New York City and the United States. Game stories varied, from hunting down ISIS terrorists in quasi-realistic but unnamed Middle Eastern settings (as in early levels of *Call of Duty 4: Modern Warfare*) to warding off terroristic aliens in outer space. Science fiction titles transposed 9/11 to inhospitable other-worlds. The War on Terror seemed capable of emerging on all kinds of fields; the threat seemed anywhere and everywhere. American history itself became a setting. EA's *Medal of Honor: Rising Son* (2003) transformed Pearl Harbor into a thinly disguised 9/11 scenario. The title presented the two disasters as essentially the same: surprise attacks on American soil marked by planes, fires, and mass panic. As Lynn Spigel noted about television culture after 9/11, "a shared and, above all moral, past" had been annexed for deployment.[5] The two national tragedies fused into one alluring game world.

While games expanded the geographic parameters of the War on Terror, they remained dedicated to a singular story of American victory. A typical first-person shooter narrative entailed a graphic display of carnage, a reason for retribution, a process of infiltrating the enemy, a series of skirmishes, and a final, triumphant fight with an end boss. Missions consisted of clearing territories of enemies and rescuing valuable figures. Progress was measured in a series of sequential levels based on extensive military operations, with victory over the enemy never in real doubt. Virtual war removed all the challenging elements of real war. Death, trauma, physicality, and political repercussions were absent from the game code, with players returning from the dead, unaffected by conflict. Most titles offered players only the "good side" of the War on Terror.[6]

In most titles, the gamer assumed the role of a tough male soldier. Replacing the valorous image of the New York firefighter, titles codified a new hero of the post-9/11 world: the patriotic soldier. The hero archetype shifted from life saver to brutal killer. The role suited the genre of military shooters perfectly. The military shooter had its origins in simple Western games such as *Gun Fight*, along with the arcade game *Battlezone* (1980), a vector graphic–based tank simulator that Atari adapted on request of the US Army for training gunners in Bradley fighting vehicles in the 1980s. It also emerged from film-inspired games such as Sega's *Rambo: First Blood Part II* (1986) and Capcom's *Commando* (1985) that forwarded the idea of the lone hero facing thousands of enemies (often within the broader context of the American soldier vanquishing the ghost of Vietnam). Post-9/11 military games *Call of Duty: Advanced Warfare* (2014) and *Battlefield 3* (2011) focused on the role of the individual against a vast but generic army. Full of graphic violence, with progress often dependent on high kill rates, such games tied nation saving with mass killing. Playing as tough male soldiers, gamers imagined being on the front line of the War on Terror. Titles such as *Call of Duty* promoted a form of virtual masculinity heavily dependent on firearm skills, bravado, and aggressive attitude.[7]

The post-9/11 enemy appeared far from complicated or fleshed out. Soldiers mostly faced an ill-sketched stereotype as enemy. As Matthew Payne observed, "the typical enemy is a non-white 'Other' who speaks a different language and who worships a different god."[8] Attempts to tackle deeper terrorist characterization or motivation mostly failed. Prior to 9/11, terrorists in games came from a variety of backgrounds, with a variety of ways of tackling them. The two-player strategy game *Terrorist* (1980) for the Apple II computer, one of the earliest titles to engage with the topic, situated terrorism as primarily an issue of negotiation. Designed by Steve Pederson, the education title explored how to move beyond the paradigm of confrontation. While one player chose to represent a government from a range of nations (including the United States, programmed and identified as strong in "individual rights"), the other player picked a terrorist group (the closest to Al-Qaeda being the National Fundamentalist Army). A sense of logic dominated game procedure: the terrorist drew up a list of demands, then employed a range of set scenarios, including building capture, hostage seizing, airplane taking, or nuclear blackmail. The government responded from its own set of reprisal techniques. Pre-9/11 gamic terrorism mostly bypassed issues of religion and sacrifice, with the aim being to avoid outright war.

In post-9/11 gaming, most titles focused on dispatching rather than negotiating with terrorists. In-game terrorist characters were typically uncommunicative. Lacking facial expression and often sporting headwear such as a balaclava, the enemy always appeared generic and without personality, rather than as an individual with a background. There seemed no common ground or reconciliation possible with the enemy; the player only had the options to shoot or be killed. The real terrorists of 9/11 provided game designers with the ideal villains. Programmers had no need for backstory. Simply identifying a character as a terrorist made them an emotionally loaded target. David Annandale noted how "the perfect enemy [has] been found."[9] A collusion between game code and cultural stereotypes blossomed. These codes and subroutines created batches of generic enemies to face, with little need for detailed (and code-heavy) characterization. In-game terrorists were all the same. Coding meanwhile helped separate the heroes from the villains. The two groups looked and behaved quite differently from each other. Popular titles maintained a welcome psychological distance between soldier and enemy. The binary nature of games suited the binary nature of the War on Terror.

Those games that sought to subvert convention and explore the mentality of the Other were controversial. *Call of Duty: Modern Warfare 2*

(2009) featured a simulation in which the player acted as a terrorist, in a level nicknamed "No Russian" and designed by Mohammed Alavi. The player, as embedded CIA operative Joseph Allen, joined a group of Russian terrorists (Inner Circle) led by Vladimir Makarov. As part of the group, the character entered Moscow Airport and massacred civilians and security forces inside. Encouraged by his team to shoot innocent people at will and create a scene of mass panic, the gamer played the role of terrorist, witnessing the bloodshed first-hand. With bloody corpses littering the departure lounge and sirens wailing, the level amounted to a "Russian 9/11." As the game's lead writer Jesse Stern explained, the level was designed to provoke discomfort and reflection: "our intention was to put you as close as possible to atrocity."[10] The level proved to be highly offensive. The *Salt Lake Tribune* deemed *Call of Duty* the most controversial game of the year, while in the United Kingdom, the violence of the level was debated in Parliament.[11]

Marcus Schulzke noted the broader gamic opportunity of playing the terrorist in video games. He forwarded the notion of a "perspectival shift" that "raises the possibility that games might offer a humanizing look at the enemies in the War on Terror and that they might give players greater insight into terrorists' motives and ideologies." Avatars and virtual selves had the potential to help understand the Other. Instead, mainstream entertainment media largely enforced a familiar stereotype of the enemy as a group of religious fanatics who engaged in senseless and unexplainable acts of violence. Games then exported such imagery to players across the world. Gamic reality thus endorsed existing and conventional understandings of external reality. As Schulzke argued, "the games confirm the overarching War on Terror narrative that terrorists are irrational and evil enemies who are unworthy of respect."[12]

Video games thus presented a War on Terror marked by simple heroes and sadistic villains. Games depicted a nation under attack, but thanks to the player's skill, the nation was capable of overcoming all odds and emerging victorious. As virtual soldiers, gamers fought for good old American values, protecting their brethren in violent conflicts and dangerous war zones. Gamic America provided an alluring fight for American freedom and democracy. According to Matthew Payne, post-9/11 shooters indulged in "nationalistic power fantasies" that portrayed the War on Terror as "a grave but necessary and patriotic undertaking." More critically, Payne suggested that such titles served a didactic function to "normalize combat spectacle as entertainment in the service of supporting the American empire."[13]

Such versions of gamic America resembled President George W. Bush's own view of America in the early 2000s and the actual War on Terror. Digital campaigns against the Taliban resembled real campaigns in Afghanistan. Individual titles promoted a take on 9/11 and the War on Terror very much in line with that of the Bush administration. In late 2001, Bush advisors met with a range of media executives to seek their backing for his policies, while internationally the administration sought allies in the war ahead.[14] Bush envisaged a comprehensive "coalition of the willing" in the War on Terror that crossed nations, politics, and media. Game companies and, in turn, game players, became unofficial and unrecognized supporters of the coalition. Blockbuster shooters such as *Call of Duty* forwarded the notion of a just and winnable War on Terror, to the extent that Ed Halter, writing for the *Village Voice*, labelled them "the next generation of wartime propaganda."[15] For Annandale, first-person shooters represented "ideological allies" to the administration; some, such as *America's Army*, were even produced in cooperation with the US military.[16] Games used rhetoric similar to that of the government. Their presentation of war as a contest between simplified heroes and villains amounted to a gamic version of Bush's Congressional address on September 20, 2001, when he argued, "You are either with us, or with the Terrorists."[17] The binary nature of Bush's America fit with the binary form of computer games.

Bush's go-get 'em attitude seemed ideal fodder for first-person shooters. Players carried out military-style attacks on a virtual ISIS. Free from the chains of liberal democracy, players did what was necessary in the fight against terror. As in other media, the Bush idea of the "just war" proliferated. Film scholar Wheeler Winston Dixon notes how, out of the ashes of 9/11, emerged "a renewed audience appetite for narratives of conflict . . . centered on a desire to replicate the idea of the 'just war.'"[18] Both popular films and video games pursued the same theme of "frontier justice." Games acted out a popular fantasy of warfare. As Annandale observes, "in the cheerleading games, then, the player can take satisfaction in dishing out the kind of fantasy justice that the Bush Administration promised but could never deliver."[19] Nick Robinson contends that games portrayed "the USA as an innocent victim of violence, so justifying a military response unbound by international norms and law."

Games had a neo-conservative payback ideology. Digital fighters employed various firearms, surveillance drones, and torture techniques. Gamers united against the invisible Other, serving as valiant soldiers in the fight against the enemy. "Upholding national values through the kinds of secret military action argued for by the Bush Administration," gamers operated as secret agents of the government. Games endorsed the myth of "winning" the War on Terror, as if the actual war itself was a simple game.[20]

Payne argues, "There is no entertainment genre that more vividly and viscerally explores the cultural values central to the US's political imaginary than the 'military shooter' produced after the September 11, 2001, terrorist attacks." Games marked the resurgence of traditional values following 9/11. Where *Metal Gear Solid* had portrayed moral ambiguity (with the true enemy in *MGS 2: Sons of Liberty* being revealed as a group calling themselves "Patriots"), post-9/11 games consistently abided by binaries and moral clarity. For Payne, the military shooter operated as fundamentally "reactionary media." Rather than create a fresh space for new ideas, 9/11 game worlds instead offered a return to conservatism. Flawed notions of a "just war" proliferated in gaming circles. Titles such as *Call of Duty* often promoted a binary division of "us" versus "them," and endorsed problematic narratives of torture and enemy capture. Video games thus encouraged an embracement of military values in mass culture. For Payne, such constant "ludic warring attempts to resurrect a virile, militarized national identity that rises phoenix-like from the ashes of the Twin Towers."[21]

Games also helped facilitate the reassertion of American symbolic power. The attacks on 9/11 had punctured the dream of America as a monolithic superpower. The disaster highlighted the vulnerability of American security and leadership. After the attacks, the Bush administration sought to "revitalize" America through a series of military set pieces and media-friendly rhetoric, drawing on American myth and mythic language to heal the wounds of sacrifice and death. The "Wanted Dead or Alive" rhetoric re-created the Wild West (itself a fantasy) as a "resurrected" fable for the immediate post-9/11 scenario. Bush's presentation of "good versus evil" resurrected the Reagan-era depiction of the Soviet Union as the "evil empire" (another effective rhetorical fantasy). Games played a role in this new rhetorical offensive. They recycled the rhetoric and promoted the resurrection of "America first." Games preached a familiar kind of American exceptionalism: the cowboy nation, threatened by Indians, but with innate and righteous superiority (and manifest destiny) over the enemy. Like the mythical gunslingers of the Wild West, gamers led America to victory by using a policy of "shoot first" and embracing an itchy (gamic) trigger finger. Titles such as *America's Army* melded patriotic endeavor with military duty. *Call of Duty* preached the real call of duty to vanquish all sense of American vulnerability through a massive multiplayer takedown of enemies. Mainstream gaming thus contributed to a broader cultural and political reassertion of American exceptionalism. The most popular titles provided a fantasy of success, wiping out all opposition—not just the opposition of the terrorists themselves, but any alternative ideas to the reassertion of myth. Post-9/11 shooters thus played a true restorative function by situating 9/11 as the prelude to the real story of restoring American values.[22]

War on Terror games equally offered something cathartic for the player. Games connected virtual selves to greater narrative arcs of crusade and redemption. They offered individuals the chance to make a difference after 9/11 and to play a role in rebuilding the nation. After helplessly witnessing the tragedy of 9/11, gamers welcomed the chance to take control and to lead America again. They became virtual heroes, and together they made up a virtual army fighting the War on Terror. First-person shooters thus played a valuable psychological role by empowering people. For Payne, "the military shooter functions as a kind of ludic antidote to the meditated 'shock and awe' of the 9/11 attacks."[23] Such games fulfilled a mass desire for a lost "victory culture" that had once been part of America.

Video games granted players the valuable illusion of "doing something." In the 1950s, in response to the nuclear threat, the Federal Civil Defense Administration embarked on a series of drills and bomb shelters, in part to organize the nation if a Soviet attack came, but also to engage the American public itself in doing something. Fifty years on, video game worlds provided citizens with the same feelings of distraction and protection. With no need to go outside, "protecting yourself" from terrorism came in the form of a click of a button, a thumb action, and a big television. Increase in television size and improved technology since 9/11 corresponded with a growing immersiveness in post-9/11 fantasies. Games provided cathartic sensations of empowerment and taking charge. Rather than building a bomb shelter, gamers proactively wiped out virtual terrorists. Players took on an action role akin to film heroes such as Rambo. Military video games suggested that the American public had a role in a war otherwise fought by contractors and soldiers. Every citizen had a part to play. As David Leonard noted, "Americans of all ages are thus able to participate collectively in the War on Terror."[24]

Through their digital play, gamers fought off terrorists, restored national pride, and reclaimed a pre-9/11 superpower status for the United States. Video games granted players the illusion of America winning the War on Terror thanks to their endeavors. In stark contrast to the complicated conflicts playing out around the world, players simply took control of America's destiny. They enacted a decisive victory over virtual terrorism. That gamers faced a digital enemy, rather than a physical one, mattered little when ultimately striving to protect a symbol and image: America the beautiful. By overcoming virtual terrorism, players rebuilt a positive and powerful image of America temporarily stricken down. The attacks of 9/11 tested the mettle of Americans, and rebuilding the nation took time, on physical, cultural, and digital levels. Like many songs, films, books, and advertisements, games asserted a narrative of unity: "Together We Stand." Video games assisted in the cultural resolve of a nation.

The Military–Entertainment Complex

> Will computers win the war on terrorism?
> MICHAEL MEDVED, "Zap! Have Fun and Help
> Defeat Terrorists, Too," 15A

Winning the virtual War on Terror sometimes overlapped with the real-life campaign. In the twenty-first century, military conflict increasingly entailed the deployment of drones and guided missiles, along with the use of remote view screens and computer simulations. Video games themselves were co-opted into the US war machine. A technological blurring marked the period, with sometimes little discernible difference between a game simulating war, and a war resembling a game. In 1996, the US military's Marine Corps Modeling and Simulation Management Office (MCMSMO) modified the first-person shooter *Doom*. Lieutenant Colonel Rick Eisiminger explained, "We were tasked with looking at commercial off-the-shelf computer games that might teach an appreciation for the art and science of war" and to "use nontraditional ideas for keeping its soldiers sharp."[25] Released on July 4, 2002, *America's Army* for the PC highlighted the growing convergence between real and virtual armies. Designed as an official recruitment tool for the US Army at a cost of $7 million, *America's Army*, a first-person shooter simulating life on the frontline as an American soldier, reflected the emerging symbiosis between gaming, military training, and patriotic duty. The PC title marked the rise of the new military–entertainment complex. The July 4 release date gave an early clue to the patriotic tone of the title. *GamesRadar* noted the timely release, on "the first Independence Day after 9/11, when we needed guidance and support."[26] The game reflected a desire to attract new, tech-savvy recruits at a cheap cost. As the *Washington Post* put it, "The Army's flagging recruitment numbers are serious business. So Army officials increasingly are turning to a game for help."[27] For the consumer, the game cost nothing. A 2003 game advertisement declared, "Citizens. Countries. Video Games. The US Army keeps them all free." By 2005, the game had more than six million users, with studies finding it to be more effective at recruiting "than all other forms of Army advertising combined."

America's Army served as a post-9/11 symbolic text and a practical recruitment tool. It amounted to twenty-first-century nationalist war propaganda and a gamic update of the 1917 "Uncle Sam Wants You" recruitment poster. As a training simulator for conflict, *America's Army* promised a "virtual soldier experience" that taught American gamers how to fight. The title molded players into soldiers. The game generated a sense of realism through its immersive soundscape (including team orders) and the care needed to avoid getting killed. It taught players teamwork, effective targeting, survival techniques, and how to eliminate terrorists on the battlefield. Intriguingly, the software divided players into two teams, who both self-identified as Americans, but saw each other as terrorists on-screen. This image swap followed the tradition of real war games, in which soldiers played sides in battle simulations. However, in practice, *America's Army* pitted Americans against Americans. Other than the computer-generated masking, the game presented terrorists as no different from patriots (with the exception of more facial hair). Tactics, firearms, and play options remained the same.[28]

America's Army highlighted in a very public way the partnership between the military–industrial complex and media–entertainment forces in the post-9/11 period. Roger Stahl described a new synergy between the military and gaming.[29] The collapse of the World Trade Center arguably rang the bell on old-fashioned conflict and ushered in a new era of high-technology, gamic warfare. However, the ties between the US military and computer games significantly pre-dated 9/11, with *Spacewar!* and *Battlezone* just two examples of historic convergence. With their focus on combat and killing, video games had always been natural allies to military war games. Similar to the "premonition titles" that foresaw an attack on New York, military-themed titles explored terrorist campaigns years prior to the official War on Terror. They offered a form of "predictive combat." The 1990s *Desert Strike* series, along with Sega's *Gunblade NY* (1995), in which the player used a Special Forces helicopter to eliminate enemies of Manhattan, depicted a range of military scenarios involving terrorism. In their fashioning of terror plots and military retribution, games, once again, explored events surrounding 9/11 prior to 9/11. This sense of gamic perceptiveness contrasted with criticism of the actual US military and its failure to plan ahead. Reflecting on the inability of official war gaming to predict and tackle contemporary conflict, Paul Virilio exclaimed, "We are always one war behind." By contrast, video games always seemed one war ahead.[30]

With ties between the military and gaming strengthening after 9/11, a number of titles sought to accurately replicate specific campaigns of the War on Terror. In 2004, Kuma Reality Games released the free military shooter *Kuma\War*, which uploaded Department of Defense data and news reports to the game engine in order to simulate actual US-led military missions. The mission to kill Osama Bin Laden in May 2011 was replicated as a game level within weeks. The Hunt for Osama, one of the pivotal news stories of post-9/11, was replicated across a number of other titles as well, including *Fugitive Hunter: War on Terror* (2004) and *Postal II* (2003). As with other forms of popular culture, video games molded the public perception of the War on Terror. For Payne, the "experiential realism" created by military shooters was a potent form of immersion and inevitably shaped views of real war for the player.[31] For Stahl, software companies held particular power in wartime imaginings, claiming that "the media paradigm by which we understand war is increasingly the video game."[32] Games also took an aggressive line on questionable wartime tactics. Post-9/11 games included torture; in *Tom*

Clancy's Splinter Cell: Conviction (2010), the player had the option of employing aggressive interrogation techniques, while in *Call of Duty: Black Ops* (2010), a captive only reveals information after being punched and forced to eat broken glass.[33]

While games replicated real-time conflict with increasing accuracy, the advent of computer-guided missiles and automated drones highlighted the extent to which "real war" was increasingly played like a computer game. In February 1991, Commander-in-Chief of the US Army General Norman Schwarzkopf felt the need to highlight that the military take-over of Iraq "is not a video game."[34] At the same time, philosopher Jean Baudrillard came to exactly the opposite conclusion, describing Operation Desert Storm as fundamentally a visually spectacular game.[35] For Slavoj Žižek, war seemed increasingly gamic: "Just as we drink beer without alcohol or coffee without caffeine, we are now getting war deprived of its substance—a virtual war fought behind computer screens, a war experienced by its participants as a video game, a war with no casualties (on our side, at least)."[36] The similarity went beyond common use of computer screens and joysticks. Wartime propaganda (of "infinite justice") and wartime operations with names such as "Enduring Freedom" increasingly resembled the tone of video games (or vice versa). As W. J. T. Mitchell noted, "the War on Terror seemed to promise an endless supply of faceless warriors, massed for interminable combat," which matched perfectly with experiences on titles such as *America's Army*.[37]

The similarity of real war with gamic war marked a growing liminality between conflict and entertainment. The US Army stressed the realism of *America's Army*, which was little different from other military shooters. Stahl referred to the advent of the "virtual citizen-soldier," a player caught in the closing gap between civilian and military identities, a phenomenon indicative of a wider militarization of popular culture.[38] Just like an entertainment television program or video game, war was something that played out on screen. Matt Delmont highlighted "the continuing centrality of visual culture to the war on terror," a conflict "mediated through an overwhelming array of visual forms and media."[39] Conventional binaries of real and imaginary no longer applied. War operated as a visual spectacle. For Mitchell, "Modern warfare is often portrayed as a de-realized spectacle, a mere simulacrum on the order of a videogame."[40] One of many cultural productions linked with the new Image/War, video games provided welcome stories of America winning the conflict, with players enrolled as dutiful citizen-soldiers. As with the terrorists that used *Microsoft Flight Simulator* to practice a final approach to the World Trade Center, soldiers trained on *America's Army* to take down Bin Laden.

As 9/11 functioned as an image/event, a symbol of America destroyed, so too did the War on Terror play out as a series of conflicts about images and symbols. In a digitally framed world, the fall of the Twin Towers was referenced by similar iconoclasm, such as the fall of Saddam Hussein's statue in Baghdad. The War on Terror was fought through symbols and imagery, as much as through conventional weapons of war. Games played their own role in this "war of images." By constantly delivering fresh images of dead terrorists, hoisted American flags, and victorious US troops, video games contributed to the larger Image/War campaign. They provided endless repetitions of failing terrorists to counter the original visual repetition of the Twin Towers falling. While 9/11 provided a symbolic defeat and setback for the United States, video games helped provide symbolic victories.[41]

Divergence from the Patriotic War

Video games also, on occasion, offered alternative screenings on the War on Terror. Not all 9/11 games followed the patriotic model set down by *America's Army*. More sophisticated titles explored some of the costs, controversies, drama, and politics of the War on Terror era. Games exposed the moral ambiguities of conflict. Hideo Kojima's *Metal Gear Solid* series continued to probe issues of trust, responsibility, and morality, as well as deeper philosophical questions of gene-enhanced warfare, identity, and self-worth in the post-9/11 era. Inspired by Joseph Conrad's novel *Heart of Darkness* (1899) and Francis Ford Coppola's film *Apocalypse Now* (1979), Yager Development's *Spec Ops: The Line* (2012) projected an even darker view of conflict. *Line* featured a range of visually subversive qualities, from the presence of inverted flags to its broader depiction of a broken civilization. Payne noted the "allegorical proxies" used in the game, with the Dubai setting substituting for the United States.[42] The title deliberately undermined gamic conventions of player honor and success, including involving the gamer, as Captain Martin Walker, in the slaughter of American soldiers beneath the flag in an early scene. Implicated in the massacre, the title encouraged the player, as Walker, to confront notions of guilt, trauma, and the misuse of violence. Gameplay also actively encouraged a mocking of the simple American hero archetype, with Colonel John Konrad taunting Walker, "Do you feel like a hero yet?" *Line* was a commercial failure, arguably because of its alternate reading of warfare.

The non-commercial game *September 12th* (2003) critiqued established thought on the War on Terror, but used the perspective of a toy-like isometric game world. Located in a Middle Eastern village, players experienced first-hand the escalation of warfare, with bombs routinely killing civilians. Game designer Gonzalo Frasca situated the game as a protest piece; Games for Change described it as a "newsgame" that "conveys a timeless maxim: violence begets more violence."[43] Taking place just one day after 9/11, the title highlighted the War on Terror as an ill-considered, retributive act. With the player caught in an endless and frustrating conflict, the game offered no clear winning strategy. Fassone described *September 12th* as "an unwinnable game about defending one of the symbols of the United States [that] certainly projects a bleak shadow."[44] Video games thus highlighted the potential for the War on Terror to continue without resolution or decisive victory. Repetitious images of war seemed most likely to replace the cyclical images of 9/11.

Virtual Shadows

Game journalist Rick Lane once claimed that "the response to 9/11 by the games industry over the past decade highlights gaming's standing within society as a mode of social commentary—it doesn't have one."[45] While Edward Said saw 9/11 as a complicated event (and certainly not a "clash of civilizations") and Jean Baudrillard noted how "the unbounded expansion of globalization creates the conditions for its own destruction," games rarely offered similar levels of deep philosophical reflection.[46] Few video games tackled 9/11 directly, most hiding behind simple allegorical commentary. However, 9/11 sent virtual ripples out into gamic America. As Annandale contends, "like other forms of popular culture, video games have mirrored, tracked, and questioned the dreams and nightmares that have shaken the American psyche in the wake of 9/11."[47] Game spaces served as experimental zones, where fantasies surrounding 9/11 and its aftermath played out. Social commentary existed, but it needed to be decoded.

Most games structured the meaning of both 9/11 and the War on Terror as a conflict between good and evil. Software companies dressed military action in neo-conservative rhetoric, with the player fighting a patriotic and wholly justified war. Many commercial titles promoted militarized violence as the only solution to world events. Only on rare occasions did the software industry challenge its own singular, gung-ho narrative. Film scholar Rebecca Bell-Metereau contends, "Movies that glorify violence or military reaction probably affect people's attitudes toward war more than they realize."[48] The same observation applies to games.

Games distracted us from the reality of the situation. They promoted a digital facade: the mass illusion of power. Military shooters granted the player the thrill of seeming to win the War on Terror single-handed, despite the computer itself deciding the variables and fundamentally shaping the experience. In their potent illusions of computer-generated godly power, video games enact Žižek's ideas about the facade of freedom and the reality of forced choice. Žižek felt that terrorism awakened Americans (and the world's citizens) "from our numbness" and, in the power of the event, offered a momentary glimpse of "the real."[49] However, Žižek asserted that most people preferred to see 9/11 on screen as a catastrophe movie. The filmic image was more alluring. Žižek compared it to snuff porn, in that people felt the "thrill of the Real as the ultimate 'effect.'" Americans interpreted events on their television as their "ultimate American paranoiac fantasy." By their very nature, games continued to pull people away from the real and back to the superficial in the post-9/11 period. Gamers avoided looking at the reality of the situation, instead letting computers guide them to more pleasant scenarios and outcomes.

Thanks to replay technology and camera phones, 9/11 was a twenty-first-century digital event as much as a physical event. It is now memorialized and revisited in countless ways. Alongside the twin reflecting pools where the Twin Towers once stood, enclosed dark granite monuments to that day, games remain "active monuments" to 9/11 and the War on Terror. The static structures marking Ground Zero tell of heroes, strength, and sacrifice; similar stories are often told through films and games. Both concrete and digital monuments serve as guides to the past and to the future: informing us about what to mourn, what to celebrate, and what direction to take.

The dark granite pools at Ground Zero serve to preserve and protect a specific memory and image of 9/11. Enshrined, overseen, patrolled, and protected, the 9/11 memorial tells a singular, official story of sacrifice and loss. In its sense of play and the provision of multiple routes, the ludic, by definition, poses an underlying threat to any one image. While many titles initially supported the rhetoric of the War on Terror, games continue to play with the experience of 9/11 and the conflicts that followed. Like Hollywood and literature, they have the potential to challenge and reshape understandings, but only if accepted by popular culture. When the Walt Disney Company volunteered to construct an American History theme park in Virginia in 1993–1994, the project was rejected on the grounds that an entertainment company could not to be trusted with something as serious as national history. Equally, when the Smithsonian in 1995 planned an exhibition of the Enola Gay that highlighted multiple interpretations of the decision to drop the bomb, curators were accused of playing with history. Video games actively play with the way we see the world, and they continue to cast multiple virtual shadows, but only if we encourage them.[50]

6

Grand Theft Los Angeles

IN 1997, a small British studio, DMA Design, released a top-down, two-dimensional video game based on driving in an urban environment while engaging in criminal activity. The title of the game came from the chief activity—stealing cars, an offense listed as Grand Theft Auto. Similar to *Death Race*, which was released twenty years earlier, *Grand Theft Auto* (*GTA*) courted significant controversy by rewarding players for knocking down pedestrians. In 2001, Rockstar North (the progeny of DMA) developed *GTA* into a three-dimensional adventure. Alongside the basic structure of driving and shooting, the software company added a range of distractions and activities, from getting haircuts to visiting strip clubs. The open-world template was immensely successful. Released in 2013, Rockstar's *GTA V* amassed worldwide sales in excess of 95 million units and $6 billion in revenue by May 2018. With total sales for the series exceeding 250 million units, the success and reach of *Grand Theft Auto* dwarfed most major film and book franchises.[1]

As in Rockstar North's other titles, *Red Dead Redemption* and *L.A. Noire*, *Grand Theft Auto* offered a sense of place and atmosphere unrivaled on home consoles. By the 2000s, console technology had improved dramatically from the Atari VCS, to the point of nearing aesthetic realism. *Grand Theft Auto V* offered one of the most detailed explorations of US life available in video games. Rockstar North spent four years developing *GTA V* at an estimated cost of around $265 million.[2] The company employed close to 1,000 staff working on elements that included detailed modeling of US architecture, replication of freeway maps, and the behavioral simulation of gang culture. Rockstar offered the closest a millennial generation could get to experiencing a virtual America. When asked in 2015 if the series could ever be set elsewhere, creative director Dan Houser replied, "My own personal feeling is *GTA* is America."[3]

Rockstar North designed *GTA V* around Los Santos, a fictional Southern California city closely resembling Los Angeles. *GTA* titles had already featured fictional versions of other California cities (alongside recreations of Miami, London, and New York City). The original *Grand Theft Auto* featured an amalgamated version of California, while *GTA: San Andreas* (2004) presented a West Coast adventure set in the 1990s. The city of Los Angeles was an ideal canvas for the series. A city widely associated with car culture, film sets, and action scenes fused with a format similarly focused on visuals and speed. David Ulin wrote, "Cars and movies . . . are the essential icons of Los Angeles, emblems of speed and light and movement."[4] Los Angeles appeared to be a city given meaning and structure by the spectacle of movement. The first film depiction of Los Angeles, a 25-second clip of Spring Street Downtown, filmed by Frederick Blechynden for Thomas Edison in 1897, captured the kinetic energy of early Angeleno street life: a world of electric trolleys, bicycles, horse-drawn carriages, and bustling pedestrians.[5] The film short showed Los Angeles as all men, movement, and action. Then came the automobile and its total invasion, dictating city conduits and the shape of a burgeoning metropolis. Writing in the 1950s, architect Reyner Banham described Los Angeles as an all-embracing "Autopia" and observed that "the freeway system in its totality is now a single comprehensible place, a coherent state of mind, a complete way of life."[6] Author Joan Didion reflected in her essay "Pacific Distances" that "a good part of any day in Los Angeles is spent driving, alone through streets devoid of meaning to the driver, which is one reason the place exhilarates some people, and floods others with an amorphous unease."[7] Cars and highways dominated the Angeleno state of mind.

When it came to constructing *GTA*'s version of Los Angeles, Rockstar digitally rendered freeways across a whole game space and structured the game world by its highway connections. Rockstar created a palpable sense of both Banham's utopia and Didion's driving alone. Across a vast virtual metropolis, players could easily spend hours driving between destinations, following road maps, speeding past police, and glancing at the Pacific Ocean. *GTA V* offered a ludic experience based on the sensation of driving in Southern California, of moving between downtown Los Angeles, California Highway 1, and Hollywood. The centrality of the automobile in Angeleno life perfectly suited the *GTA* franchise's own focus on reckless driving and auto theft. Expensive cars could be located in the Hills, while respray garages were located on the edges of town. As Ian Bogost and Dan Klainbaum noted, the "series title seems particularly suited for Los Angeles, a city famous for its carjackings, freeway shootings, high (and low)-speed chases, and its generally unrivaled car culture."[8]

In the early twentieth century, lured by the distinctive light of the West Coast as well as its geographic distance from Thomas Edison and his attempt to copyright and control a fledgling picture industry, East Coast moviemakers moved to Los Angeles en masse. They colonized the nascent metropolis for the purpose of independent film. Writing in the 1910s, poet Vachel Lindsay pondered, "Will this land furthest west be the first to capture the inner spirit of this newest and most curious of the arts?"[9] Lindsay hoped for the "evolution" of California and film combined, "for the state and the art to acquire spiritual tradition and depth together." In its skin-deep buildings, re-appropriated Missionary style, and fledgling star base, Los Angeles soon took on the patina of one giant movie set. It became a place made by movie studios, for movie studios, and duly captured on screen. Twentieth-century Los Angeles became the city of image, make-believe, fakery, and facade. Michael Sorkin later called it the "most mediated town in America, nearly unviewable save through the fictive scrim of its mythologizers."[10] The filmic version of Los Angeles became unescapable.

Rockstar drew on this filmic imagery when coding its gamic version of the city. Despite Houser's insistence that *GTA* represented something new (noting how he abstained from watching all Los Angeles–based movies: "we don't need to hark back to film when technology allows us to produce our own response to real places"), Los Santos was unavoidably movie-based.[11] A digital twin to the Hollywood sign, the Vinewood sign in *GTA V* looked down across Los Santos, a filmmaker's daydream digitized. The gamic version of Los Angeles drew heavily on the imagined landscape of film and literature. The series offered a "joyride through genre cinema and literature," with artistic nods to film directors Quentin Tarantino and Michael Mann, television series such as *Lost* and *CHiPs*, and the work of literary mythologizers of Los Angeles such as Bret Easton Ellis.[12] One elongated heist scene in *GTA V* replicated verbatim Michael Mann's groundbreaking crime sequence in the movie *Heat* (1995). As traditional media forged a comprehensive image (and story) of LA across the twentieth century, that image in turn served as the template for the new media of games. As Nate Garrelts contends, video games offered "adult media culture recycled."[13] Sam Sweet for the *New Yorker* noted, "Los Santos is not a real city; it is the Los Angeles of cop dramas and heist flicks and music videos, a love letter to the synthetic Los Angeles of the imagination."[14]

Alongside the influences of Hollywood film, Rockstar tapped into the broader mythology of Los Angeles rooted in California and the West. Much like arcade games of the Wild West, games of the urban frontier reworked and recoded popular culture fragments and established mythologies of place. Rockstar programmers indulged in a range of mythologies, commenting not just on Los Angeles's enduring reputation as the place where things happen and dreams are made, but also on its reputation for seediness, superficiality, celebrity television, and consumer excess. *GTA* tracked Angeleno culture from plastic surgery to star power, from microbiotics to muscle beach. It also explored the highs and the lows of the California Dream. In the words of Mike Davis, it captured both the sunshine and the noir of the City of Angels.[15] Historically a place to escape to, Los Angeles became a game to escape in. In constructing a computer version of Los Angeles, Rockstar adopted (and subtly transformed) specific notions of place, exaggerating the already exaggerated side of Angeleno life, while dismissing much of the mundane normalcy of city living. Rejecting the missionary romance of Los Angeles depicted in classic novels such as Helen Hunt Jackson's *Ramona* (1884), Rockstar instead cast Los Angeles as chiefly a mix of gangland and Hollywood bubble.[16] Rockstar created a "glitzily superficial city of Los Santos, a warped mirror of Los Angeles," based around celebrities and crime, psychos and surf culture.[17]

In the process of building and refining Los Santos, Rockstar established its own fictive map of the city. Programmers condensed some streets, while widening others. South-Central gangland became smaller (ironic, given the game's focus on crime), while beaches and board-walks expanded. In the process of coding Los Santos, the programming team studied thousands of city maps, building plans, and historic photographs. Hundreds of city buildings re-materialized in the game world, a mass assemblage of real-life architecture digitized. The boundaries between the real and the fake inevitably blurred. Staff at Hotel Figueroa used *GTA V* images by mistake to advertise its location, while internet fan sites published "spot the difference" pictures featuring the real and digital attractions.[18]

Rockstar zoned the city around specific themes and identities. In *Los Angeles: Architecture of Four Ecologies* (1971), Banham depicted the real Los Angeles as consisting of distinct zones: Surfubia (the beach cities), the Foothills (the rich white enclaves), and the Plains of Id (down-town sections), all held together by an Autopia grid.[19] Los Santos similarly featured three zones linked together by a freeway skeleton and broadly replicated Banham's structure by highlighting areas of beaches, riches, and downtown business. The sense of Rockstar forging its own visual map and built city meanwhile was reminiscent of the efforts of artist and photographer Ed Ruscha in the mid-1960s to discover and document the real Los Angeles. In a huge range of aerial photographs and drive-by scenes, Ruscha provided his own taxonomy of the post-war West Coast metropolis.[20] Ruscha's Los Angeles was a place of placeless-ness, made up of generic, mundane, and highly commercial structures; it was also, at times, hauntingly people-less, as if the metropolis had become a concrete ghost town or a nuclear bomb site. Ruscha attempted to map the meaning of Los Angeles, to visually document and capture its essence. Likewise, Rockstar's programmers looked for the inner meaning of Los Angeles in their coding project.

For Ruscha, the essence of Los Angeles resided in its distinctive urban topography. Arguably, however, the true essence of the city derived not from automobile fumes, celluloid fixation, or street maps, but ultimately from nature: specifically, in the form of Los Angeles's light. The distinctive glow across the cityscape is the result of the San Gabriel Mountains trapping air (and pollution) inside the "bowl," with the baking sun doing the rest. The reason why moviemakers flocked to the City of Angels, the distinctive light of Southern California marked it as both a temperate paradise and an unearthly realm. From David Lynch's surrealist film projects to the showcase of lights from police and press helicopters beaming down on OJ Simpson's car at night, Los Angeles existed most where the beams shone. As the editorial team for the *New Yorker* declared, "the distinctive light of LA—the way it can cast the city in hyperreal relief or wrap it in a dreamy haze—is legendary."[21] For Mike Davis, the light of Los Angeles symbolized the distinctiveness of place, the sunshine and noir of the city marking the polar extremes of living there. In order to authentically capture Los Angeles, Rockstar programmers worked on how best to simulate the Southern California brightness— how to master the light. The resultant game light provided sublime sunrises and sunsets, but also a sense of game time. Night and day in *GTA V* worked by a computer-controlled program, the typical twenty-four-hour cycle reduced to forty-eight minutes of game time. Time-lapse photography showed digital traffic moving through downtown as the sun rose and fell over Los Santos. By perfecting the glow, Rockstar captured the core identity of Los Angeles. The day/night cycle also enabled game elements; it triggered the repeat of daily tasks, provided set intervals between missions, and helped structure the in-game stock exchange. Crime worked by the cycle. The light and dark highlighted the different tones of the game world: "Los Santos is a place of contrasts, of luxury and poverty, tranquility and violence, beauty and ugliness" games journalist Carolyn Petit noted.[22]

The lighting of Los Santos helped foster the sense of a living game world, a city with its own independent chimes and motions. Digital scholar Vince Miller talks of games in terms of vividness and sensory depth: the deeper the visceral experience, the better the game.[23] *GTA V* offered far more than aesthetic realism (or telepresence); it provided a fully immersive and emotional experience. It offered the illusion of living in an alternate, digitally formed West Coast. To relieve the boredom of driving well-greased automotive hot rods between missions, players tuned their car radios to listen to various stations, from talk radio to hard rock, or chatted with their in-game partners in crime sitting next to them. Along with individual missions and stories to follow, players could choose to take yoga classes, play tennis, or get a lap dance in their free time. They surfed an in-game internet, phoned in-game friends, and even took hallucinogenic drugs (with the game character becoming wobbly and vision blurring). Rockstar produced an interactive world above and beyond the simple actions of shooting and driving. Moreover, the city had its own trajectory independent of player action. Buses collected workers from their homes in East Los Santos, moving them from one side of the city to the other, to tend rich lawns in Rockford (read: Beverly) Hills. Characters Trevor, Michael, and Franklin seemed busy even when a player was not playing as them (the player would find each character in new circumstances when returning to them). The city dynamic extended to seemingly inexplicable events: players discovered mysterious apparitions at the haunted Blaine County Motel in Sandy Shores and pondered the function of religious cults in Chiliad Mountain State Wilderness.

This sense of digital place making afforded Los Santos a cohesion and an identity arguably more convincing than its real-life twin. Visiting Los Angeles in 1939, Christopher Isherwood wrote in his diary, "What the arriving traveller first sees are mere advertisements for a city which doesn't exist."[24] Like the folkloric cities of Atlantis and El Dorado, the City of Angels existed more as a mirage, a booster's advertisement, a myth or a daydream, than as a concrete, lived experience. Los Angeles seemed a projection of the imagination, something fundamentally ethereal and unfinished. As Adam Kirsch described the literary Los Angeles, "where you expect to find the city itself, there is only a carnival of metaphors."[25] Rockstar designed a game from this underdesigned and unfolding realm. Programmers deleted identity-less parts of Los Angeles— the excessive suburban sprawl and the sweeping gangland—and replaced them with core architectural spectacle. The city criticized for lacking a center became more navigable, focused, and meaningful in its digital form. Rockstar gave new order to the metropolis. Programmers filled in the gaps. The software company constructed its own digital Los Angeles as a living, breathing, emotional world, an alternative reality (or fantasy) for players to enter, with the promise of something more. Houser explained, "Books tell you something, movies show you something, games let you do something."[26] For the player, Los Santos offered the opportunity to both explore the surface of Los Angeles, its billboards and its beauty, and peer underneath, into its world of crime and deceit. Like a classic trope from a hard-boiled detective novel, the player wandered a realm of darkness and light, finding new truths and insights.

In a contribution for *Esquire* magazine in 1960, novelist Norman Mailer presented his sense of the essence of Los Angeles as a "land of the pretty-pretty[;] the virility is in the barbarisms, the vulgarities, it is in the huge billboards, the screamers of the neon lighting, the shouting farm-utensil colors of the gas stations and the monster drugstores, it is in the swing of the sports cars, hot rods, convertibles, Los Angeles is a city to drive in, the boulevards are wide, the traffic is nervous and fast, the radio stations play."[27] Mailer could easily have been driving in Los Santos. Mailer recognized Los Angeles as a realm caught in the moving image—"one has the feeling it was built by television sets giving orders to men." Written on the cusp of the gamic age, with *Spacewar!* just two years off, such a sentiment prefigured a time when machines would indeed build worlds and give orders to men. Mailer also declared (his) Los Angeles "a playground for mass men."

The Playground of Los Angeles

In the Red Hot Chili Pepper's rock anthem "Californication" (1999), songwriter Anthony Kiedis explored the satirical landscape of Los Angeles and the Golden State.[28] Moving through a montage of California imagery, Kiedis sang of "dreams of silver screen quotation," "celebrity skin" breaking "the spell of aging," "hardcore soft porn," teenage pregnancies, and a world where everything is for sale. The California Dream had both light and dark qualities. Singing as much about himself as about California, Kiedis stated, "Destruction leads to a very rough road, but it also breeds creation." California appeared simultaneously imagined and lost, like a playful dream.

The process of Californication also closely resembled a game. For Kiedis, both California and "the final frontier" were made in a Hollywood basement, designed and manufactured like a film, and eminently playable. The music video for "Californication" functioned as a three-dimensional video game, following the fortunes of the band as four game characters who explored the open world of California as their playground.[29] The characters snowboarded the Golden Gate Bridge, joyrided on the highway near the Hollywood sign, dined at Andy's Donuts, and grappled with a grizzly bear. The game style consciously mimicked Sega's *Crazy Taxi* as well as DMA's early sandbox games. The music video ended with a "Game Over" screen and a "Next Game" prompt. The California experience was ultimately rendered as satirical play.[30]

Rockstar's *Grand Theft Auto* series similarly presented the California experience as satirical play. *GTA V*'s Los Santos was a cornucopia of fantasy living and opulence. Publicity for *GTA V* resembled California vacation advertising, presenting the title as all about celebrity lifestyle, beaches, surfing, and beautiful people.[31] On the surface, the game world offered an opportunity to "live" the California dream, as if both programmer and player had bought into a state-bound mythology. However, like Kiedis, Rockstar satirized the experience, highlighting the setbacks and cons of Californication. The player discovered a fantastical landscape to play in, but one fractured by duplicity and disappointment. *GTA* showed the emptiness behind the beauty of Los Angeles's bodies, beach, and sand. As poet Carolyn See said of Los Angeles, "It looks good, but that's it."[32] Like the rock music video, Rockstar accentuated the fantasy architecture of Los Angeles—its giant doughnut food stops, fake missionary buildings, and movie sets—as well as the fakeness of the experience. Drawing on the historic fantasy space of Los Angeles—Hollywood, Disney, and the make-believe—Rockstar programmed a game world based on mass illusion.

Of course, the sense of California as playground, game, and mass illusion predated both Kiedis and Rockstar. Late-nineteenth-century visitors envisaged California as an escape from the industrial and confining East, and as a chance for playful reinvention. The first parks in Los Angeles owed their existence to patterns of geography: land deemed unsuitable for building led to the opening of Elysian Park (1886) and Griffith Park (1896). With the rise of Hollywood came a new sense of Los Angeles as entertainment spectacle. Carey McWilliams described early-twentieth-century Los Angeles as "essentially a tourist town. Like most tourist towns, it had its fair share of freaks, side-shows, novelties, and show-places."[33] A mix of boosters and billboards, bibles and movie bit-parts, Los Angeles seemed already a place of extremes, an Eden and a dead end, as well as a realm of deviant playfulness. Life seemed quantifiably different, verging on the unreal. Mike Davis related how in the 1940s, "To move to Lotusland is to sever connection with national reality, to lose historical and experiential footing, to surrender critical distance, and to submerge oneself in spectacle and fraud."[34] McWilliams called post-war Los Angeles "a great circus without a tent," while French intellectual Simone de Beauvoir, visiting in the late 1940s, described a "kaleidoscope" city, "a hall of mirrors," and "a huge silent fairyland."[35] Allusions to amusement and play continued to linger around the City of Angels in the latter half of the century, a filmic cousin to the emergent "adult playground" of Las Vegas. Eccentric tycoon Howard Hughes moved between the two landscapes as if remaining in a perpetual movie fantasy. Like Las Vegas, Los Angeles carried a sense of expectancy and fantasy fulfillment. By the 1990s, post-modern theorist Edward Soja conceptualized Los Angeles as "a theme park paradise."[36]

With *GTA V*, players came to Los Santos as the latest version of California to explore. One journalist observed how the game operated as a seductive "form of virtual tourism."[37] The sense of newness was initially bewildering. Los Santos behaved as a virtual tourist space and a just-finished movie set. When Fredric Jameson first entered the Westin Bonaventure Hotel in Los Angeles in 1975, a towering futuristic project melding recreation, shopping, and stay, he was both enraptured and dislocated by the experience.[38] With its endless corridors of steel, hidden entrances, and revolving restaurant, Jameson felt lost, peripheralized, but also cognizant of being on a frontier of a new kind of space. For Soja, the Bonaventure served as an excellent symbol of the post-modern city of Southern California: "its architecture recapitulates and reflects the sprawling manufactured spaces of Los Angeles."[39] Los Santos similarly represented a realm of "future space," a fusion of recreation and fantasy, and a fundamentally new geography to negotiate. *GTA V*'s own Bonaventure existed as a steely shell in the downtown section, approachable by car or foot, but, with no obvious doors or entrances, inaccessible to the player.

Los Santos encouraged discovery in the sense of exploring known places, but seeing them from a new digital perspective. Hidden in the rocky area between the Alamo Sea and the Pacific Ocean, a few miles to the north of the city, Raton Canyon provided gamers with a beautiful place to rest and admire the digital wilderness. Raton Canyon owed its existence to a fictional canyon featured in beatnik author Jack Kerouac's novel *Big Sur* (1962). Kerouac likely created Raton as a reference to bookseller Lawrence Ferlinghetti's Bixby Canyon, a property that he won in a gamble. As the narrator in *Big Sur*, Kerouac visited Raton for a period of relaxation. The *GTA* player "visited" Kerouac's world, exploring the author's imagined spaces. One gamer explained, "I love the Jack Kerouac references here. The area feels like an ode to the Big Sur."[40] Rockstar digitized both real and imagined spaces for players to explore, and added new levels of interaction to mythic spaces.

Rockstar also fashioned Los Angeles into a theme park. Norman

Mailer wrote in 1960 of the Los Angeles experience as "here to live, or try to live in the rootless pleasure world of an adult child."[41] Similarly, gamic Los Angeles focused less on the real-life issues of work or real estate, and more on the hope of an adult-child experience based around adventure. *GTA* drew heavily on "theme park Los Angeles," a tourist space knotted together by Knott's Berry Farm, Pacific Park, Chutes Park, Universal Studios, and Disneyland. Such amusement parks provided recreational space, a range of activities and games, and an overarching theme of fantasy. The game world of *Grand Theft Auto* did much the same. Players could skydive over the Pacific Ocean. They could ride a rollercoaster or a Ferris wheel at Pacific Pier (a facsimile of Santa Monica's pier). With its list of attractions and entertainment coordinates, the player's onscreen Los Santos map resembled a visitor's theme park brochure. A navigation sidebar highlighted the range of activities on offer, from tennis and golf to night clubs and gun clubs. The edges of the game world resembled the outer berms of theme parks, while fake buildings resembled the visual but inaccessible architecture of Disney's Main Street. In the same way that amusement parks offered simulated risk—the feeling, but not the reality, of peril—*GTA* offered a sense of danger, but from the safety of the armchair and a gamepad. Car crashes happened without producing whiplash, smog lines induced no asthma, and shootings caused no pain.

Like America's first "people's playground"—Coney Island, New York—*Grand Theft Auto* provided an outlet for deviance, a "magic circle" to play in, with few of the real-world costs. Nicknamed Sodom by the Sea, Coney garnered a reputation in the early 1900s for seediness and inappropriate play, and duly invited moral panic. *GTA* seemed a kind of digital successor, offering a virtual playground of dubious joys, and similarly became vilified by politicians and professionals alike. Republican Congressman Fred Upton (Michigan) accused Rockstar of disregarding rules to "peddle sexually explicit material to our youth," while activist lawyer Jack Thomson called *GTA* a "murder simulator."[42] For Gaines Hubbell, *GTA V* resembled "a satirical carnival," rotating "around grotesque realism" and highlighting "the seediness of life."[43] Coney Island also functioned as a parody landscape, full of caricatures and flamboyant characters, willing to play with the audience. At Coney's Insanitarium, punters finished a steeplechase ride to find themselves exiting onto a public stage, as part of a circus-like performance, prodded by electric rods, blown by air vents, and forced to perform like monkeys. Coney's audience became the object of entertainment. Rockstar similarly mocked its audience by parodying their whims and desires, such as hinting at in-game jetpacks, spaceships, and even sightings of Bigfoot.

As an antidote to the sordid deviance of Coney, Walt Disney opened Disneyland in Anaheim, California, as a homage to a different sort of America, a land of wholesome family values and social conservatism. Disneyland represented a complete fantasy landscape, a bomb shelter for the nuclear family to escape the horrors of the outside world and the threats of imminent Armageddon posed by the clash of superpowers and their atomic armaments. It offered a perfect retreat to a 1950s fantasy world of domestic bliss and high-tech appliances, closely connected (both geographically and intellectually) with the greater fantasy of the California Dream. In 2001, Disney opened the California Adventure next to Disneyland, a "themed California" to complement the original vision.[44]

Fifty years after the opening of Disneyland, Rockstar provided an escape conceptually far closer to Coney Island than to Disneyland. Its digital facsimile of Los Angeles omitted Anaheim, and Rockstar shied from critiquing the litigious-prone Disney corporation. In 1997, Baltimore filmmaker Jay Kristopher Huddy produced a PC game *Los Disneys* (1997, re-released 2007), a *Doom*-style first-person rampage through a dark and twisted version of Disneyland, inhabited by killer animatronic cartoon characters and weapon-wielding kids, and featuring a bizarre plot involving a Disney doomsday device.[45] *Los Disneys* provided a shock parody of American popular culture. Rockstar did something more ambitious: it forged a satirical America for all to play in. Los Santos resembled an anti-Disney space and statement: a themed experience positioned in stark opposition to the family values and cartoon cuteness of Mickey Mouse. *GTA V* provided a land of fantasy, but one marked by dissidence and dissonance. Rather than a Tomorrowland of technological utopianism, Los Santos was about muscle cars and brute force. Rockstar turned the Tinseltown of celebrities and consumption into something much darker. A land of make-believe and virtue transformed into a playground of disbelief and urban decay. Rather than Peter Pan characters, low-bit criminals such as Franklin and Michael shaped the core narrative. Rockstar presented California as primarily an adult playground, with no children in evidence. *Grand Theft Auto V* represented a gangster's paradise far more than a family idyll. Rockstar coded its version of America as a playground of violence: a theme park of aggression, shooting, and speed. It provided a fantasy for those skeptical of the niceties of the American dream, a game for the Disney disenchanted. Los Santos provided a digital post-Disneyland. In its replication of crime, economic disparity, and racism, it arguably provided a virtual California more accurate than the California Adventure offered at Disney.

On visiting Los Angeles in the 1980s, Jean Baudrillard declared, "All of Los Angeles and the America surrounding it are no longer real, but of the order of the hyperreal and simulation."[46] Baudrillard cast Los Angeles as the prime example of a symbolic realm: a place based around imitation and image making. In many ways, Los Santos provided a twenty-first-century equivalent, a coded update. As Ben DeVane and Kurt Squire explained, "The game world itself is neither real nor fiction but hyperreal."[47] Los Santos symbolized California in the 2010s. It gathered together popular statements and slogans and threw them into a themed world. The player traversed a never-ending assortment of infomercials, clichés, and stereotypical images. As Zach Whalen explained, "Los Santos is a simulacrum of sorts, an iconic template composed of surfaces borrowed from film and media that has been glued into the ludic framework required by the game."[48] Keith Stuart from the *Guardian* noted, "*Grand Theft Auto V* is not really a game about story or mechanics, even if it wants to be—it is a game about spectacle and experience."[49] Rockstar provided a gamic America of familiar referents. Rather than perfectly replicate the Los Angeles landscape, Rockstar crafted its own trademark version of the city. As with Disneyland, Los Santos functioned as a recognized exaggeration, a colorful take on the real. At Disneyland in the 1950s, nobody was under the illusion that the rubber crocodile on the Jungle Cruise was real (and thus capable of devouring guests); rather, the appeal of the ride was that it provided the playful, harmless, Disney version of the "jungle." Just as Umberto Eco felt disappointed on visiting the real New Orleans after a visit to Disney's facsimile, gamers admired Rockstar's take on Los Angeles.[50] With the ability to roam sea, land, and air without repercussion, some no doubt preferred gamic Los Angeles to the real Los Angeles. The fantasy of living wildly, of satirical play, pulled them in.

A Male Fantasy

In the Chuck Palahniuk novel *Fight Club* (1996), a nameless protagonist battles insomnia and his inner demons. He exists in a nihilistic humdrum existence as a car parts locater. He wanders self-help groups, from Alcoholics Anonymous to cancer survivors, looking for assistance and meaning. He feigns various life-threatening illnesses to feel healthy. He ultimately finds salvation in the character of Tyler Durden and a male-only space designed for violence: a secret underground organization known simply as Fight Club. Palahniuk's novel (and the corresponding David Fincher movie) captured a growing sense of sterility and emasculation in 1990s male culture. Cognizant of contemporary literature providing women with new direction, Palahniuk wrote with the hope of creating a "novel that presented a new social model for men to share their lives."[51] For readers and viewers, *Fight Club* provided a fictional space in which masculinity was explored. The novel and its film adaptation triggered broader discussions about male crisis and renewal. Societal notions of masculinity also influenced video game culture. Rockstar's *Grand Theft Auto* occupied a similar cognitive and temporal space to Palahniuk's novel. For the overwhelmingly male audience, it provided a protected space of action, violence, and freedom. It amounted to a virtual Fight Club.

Grand Theft Auto V cast Los Santos as fundamentally a male space. Rockstar designed the city around three male identities and their corresponding territories: retired Michael, residing in an expensive gated house in Rockford (or Beverly) Hills; young and poor Franklin, an African-American, struggling in South Los Santos (South-Central); and white psychopath Trevor living in Blaine County (an amalgam of border/outside LA regions). Rather than men of honor, sacrifice, and ethics, Rockstar scripted all three characters in *GTA V* as inherently troubled men. Far from superheroes or frontier blazers, the trio exhibited significant moral flaws. Taking control of Michael, Franklin, or Trevor, gamers learned to play maladjusted male roles, adjusting to Michael's sardonic disillusionment with life or Trevor's penchant for extreme acts of hostility and physicality. *GTA* thus featured a performative aspect; players learned a range of coded masculinities, albeit all men ultimately let down by the American dream and on a path of waywardness and transgression.

Grand Theft Auto functioned as a valuable psychological space in which to explore aspects of male identity. Noting how gaming technology facilitated role-play, Elena Bertozzi labelled the Rockstar series a "playground for masculinity."[52] Houser explained that "the concept of being masculine was so key to [the GTA] story."[53] In the highly persuasive *The Feminine Mystique* (1963), Betty Friedan labelled the home a "concentration camp" for women living in the 1960s. By the 1990s, some men felt society had similarly failed them. Consoles and computers offered an escape from their own "concentration camp" realities of pressured jobs and increasingly domestic roles. As early as the 1970s, computer advertising targeted the white-collar American male, representing the technology as vital to "productive information-age labor."[54] Games provided an expression of felt repression. Consciously or not, Rockstar provided a safe zone for exploring masculinity by forging a world unshackled from traditional rules of behavior. Like other action games, *GTA* operated as an escape valve, a digital realm to let off steam. Simple acts of virtual violence belied their cognitive depth. *GTA* captured a growing sense of male indignity, disillusion, and protest. "Enjoying the transgressive thrills of living the life of a young black hoodlum," one magazine reviewer explained his enjoyment of the game as essentially "a defiant finger up to the Man. Or more specifically, to the stifling worthiness of our modern culture." The reviewer noted how the title offered, "sheer joyous escapism into a universe where you can still act out your most politically incorrect fantasies. Gaming is the last bastion, the Helm's Deep freedom of expression."[55] For some players, *Grand Theft Auto* signified male liberation.

GTA catered to a desire for ownership over male identity. Players owned their virtual bodies as well as their virtual cars. They could personalize their in-game characters, making Michael or Franklin visit the Los Santos gym, get a tattoo, or drink excessively. Comparing baboon society to *GTA* society, Bertozzi suggested the game fulfilled innate animalistic tendencies of men (to hunt, to mark territory, etc.).[56] The Rockstar script, however, offered more sophistication than basic desires, exploring male identity in terms of adult responsibilities and psychological crisis. Presented as a troubled father, character Michael struggled with family issues. With his wife cheating, his daughter enraptured by dubious Hollywood characters, and his son addicted to drugs, Michael regularly attended therapy sessions. Taking control of Michael exposed players to the character's trials and tribulations, forcing them to engage with complicated domestic issues. Michael fretted over the sensation of someone controlling him, referencing not only the player's control, but also the character's lack of purpose in greater society. As a film buff, Michael regularly attended theaters in Los Santos. Designed to be watched simultaneously by both Michael and the player, the foreign art film *Capolavoro* told the story of a man who heard unwanted female voices in his head and was plagued by guilt over the death of a friend. Watching the in-game film encouraged the player (as Michael) to reflect on matters of self-perception, ego, and male relationships with women. A reviewer for the *Guardian* highlighted how "Rockstar wants to interrogate the relationship between player and game."[57] *GTA* offered players a quest for contemporary white male identity and at times conjured a new "male mystique" through play.

Los Santos served a range of male wants and desires. Rockstar provided a virtual city tailored to male fantasies. In-game maps denoted markers of male desire, from strip bars to car shops. From fast cars to "loose women," everything seemed within easy reach. The overwhelming sense of "take what you want" referenced both an alpha male mindset and a regional quality. Houser explained, "LA is this embodiment of 20th-century American desires: the houses, the gardens, the tans, all slightly fake."[58] "Taking it" seemed a quintessentially "LA thing." On November 5, 1913, engineer William Mulholland stood before the Los Angeles Aqueduct (with its water taken from Owens Valley) and famously offered to the citizenry a bounty of water, the new lifeblood for the metropolis. "There it is. Take it," Mulholland proclaimed.[59] As dramatized in Roman Polanski's film *Chinatown* (1974), Los Angeles boosters and criminals alike thought the same, taking all kinds of riches. In the post-war era, "taking it" applied to far more than water. In *Less Than Zero* (1985) by Bret Easton Ellis, the character Rip explained that in Los Angeles, "What's right? If you want something, you have the right to take it. If you want to do something, you have the right to do it."[60] The same core idea applied to the world of *GTA*. Beginning with the simple act of grand theft auto, players could branch out to stealing money, real estate, or lives. The whole of Los Angeles seemed there for the taking. Mullholland's water grab of the 1910s appeared accelerated to its natural conclusion.[61] Rockstar had, after all, "taken" the city—its architecture, crime, essence, water—for re-appropriation and simulation:

the team had constructed a grand theft Los Angeles. The player then stole what he wanted in the game.

Like most video games, *Grand Theft Auto* encouraged players to freely take lives, money, and resources. In *GTA*, the scale of a player's taking contributed to their overall status, their actions moving them up and down a "most wanted" list. A reference to the reality television show *America's Most Wanted*, the computer code rated gamers by the notoriety of their endeavors. Los Santos provided a fantasy land themed mostly around killing. Like Michael Crichton's *Westworld*, where male clients entered a themed world to become cowboy gunslingers and pick up prostitutes, Rockstar's *GTA* fulfilled a male fantasy of continually participating in heists and takedowns. However, like *Westworld*, *GTA* still had limits enforced by programs, maps, and subroutines. As Raymond Chandler related, "You can't have everything, even in California."[62]

In Palahniuk's novel, violence served as the modus operandi of masculinity. As Tyler Durden requested of the protagonist, "I want you to do me a favor . . . I want you to hit me as hard as you can." Violence similarly was the core processing unit of *GTA*. The world of *Grand Theft Auto* operated as one giant fight club. All three protagonists in *GTA V* showed immense physical aggression. With his psychotic element, Trevor, in particular, exhibited an unbridled, unhinged addiction to violent behavior not that dissimilar to the more deranged characters in Palahniuk's novel. Trevor could be found beating up Groove Street gang members or curb-stomping to death partners of women he had just had sex with. He embraced a code of violence without remorse. Trevor also seemed the most naturalized resident of the digital city landscape. While Michael remained disillusioned with his Hills lifestyle and Franklin questioned the logic of a criminal lifestyle (one commentator argued that Franklin "never accepts the crimes he must commit as normal"), Trevor fully embraced the widespread culture of violence around him. As Gaines Hubbell noted, "[Trevor] is morally repugnant, yet fits in perfectly in Los Santos."[63] Assuming control of such characters, the player was asked to perpetrate violent acts, to "hit me as hard as you can." Constantly caught in fistfights, gunfights, and drive-bys, players succeeded in their missions through well-honed destructive behavior. Shooting, fighting, and driving over people were the actions of the game that players needed to practice and perfect. The automobile served as a ruthless male implement as players employed it to carry out damage. As David Fine noted, ever since the 1920s, the car provided a "death instrument or metaphor for the illusory promise of mobility" in Los Angeles.[64] For Gonzalo Frasca, the emphasis on violence reflected a liberating quality of the game space; players no longer felt trapped by real-world confines or dictates. The world of *GTA* appealed precisely because "there is no need for negotiation. Car crashes, baseball bats and flamethrowers are the tools for 'communicating' in this world."[65] Players reverted to formative archetypes of the hunter and fighter. They engaged in visceral acts in a virtual space, getting the adrenal reward for fights without the physical scars of a real fight club. Tanner Higgin described the violence in GTA as "satiric and cathartic," suggestive of a conscious playfulness and self-awareness to the proceedings.[66]

Grand Theft Auto also depicted men as criminals. In *GTA V*, all three lead characters were felons. As Fine comments on the fictional world of Los Angeles crime writer James Ellroy, "there are no good guys."[67] From dime novels depicting scandalous rogues to stories of Al Capone to modern crime scene investigations, the criminal world has consistently entertained audiences. A step up from empathizing with troubled characters in films such as *The Godfather* (1972) and *Heat* (1995), video games offered people the chance to play the criminal. One of the earliest examples, *Bank Heist* (1983) for the Atari VCS, invited the player to rob banks and employ dynamite to thwart chasing cop cars. Like the *GTA* series, *Bank Heist* took place in fictitious American worlds: Heistown, Illinois; Flat Broke, Arkansas; and Bankersfield, California. Rockstar's *GTA* series introduced all manner of robberies and heists for the player to perform, from small-scale gas station holdups to complicated, team-based maneuvers. Cognizant of the appeal of playing a villainous archetype, the game invited a performative aspect. Taking control of Michael or Trevor, the gamer played a character, and acted out a Hollywood-style script. However, while *GTA* glorified criminal acts, it rarely venerated the act of *being* the criminal, often highlighting the trials and tribulations faced by its lead characters. The degree of self-identification and immersion depended on the individual person, likely linked to time spent in the game world. As one Seattle gamer explained his fascination with *GTA*, "I get to indulge my dark side a little bit, but it's more than story elements. You get attached to the character. I don't feel like I am the character, but I empathize with the character."[68]

Rockstar actively invited male transgression. Since the game's inception in 1997, the reputation of the *GTA* franchise rested in part on its ability to court controversy and test the boundaries of social acceptability. Beyond the fantasy of playing criminal (itself a significant transgression), *GTA V* included a range of behaviors all considered taboo, culturally frowned upon, or reckless. At Vanilla Unicorn (a reference to real-life strip club chain Spearmint Rhino), a strip bar on Elgin Avenue, Rockstar provided a recreational pursuit largely frowned upon by middle America. Rockstar then invited gamers to break club rules and inappropriately touch the strippers. Inside private booths, players could fondle scantily clad women when watchful security staff looked the other way. If caught, players were first warned, then forcibly ejected. In the mission "By the Book," the player, as psychopath Trevor, tortured a suspected terrorist.[69] Trevor took pleasure from the experience, in the vein of serial killer Patrick Bateman in Bret Easton Ellis's *American Psycho* (1991). As Trevor, the player received gold status for the mission only if he employed all four methods of torture. Across Los Santos, opportunities for transgression continually presented themselves, from shooting gang members, police officers, and senior citizens to simply driving over pedestrians. Restricted to a fictional game world, such transgressive behavior happened without real-world repercussions. Rockstar provided a simulation of transgression and an illusion of freedom. Arguably, *GTA* enabled a hollow power fantasy. Like Palahniuk's *Fight Club*, transgression became idealized, dream-like, and acted out. It also seemed complicated by the exact identity of the perpetrator. In *Fight Club*, the protagonist ultimately discovered that he is both himself and Tyler Durden, while in *GTA*, the player is constantly both himself and a reckless avatar.

Men joined Fight Club in Palahniuk's story primarily because they were lost, searching for meaning, and wanting to belong. Fight Club offered elite membership, secrecy, and loyalty. It was conspiratorial and cult-like, but also open to the ordinary man. *GTA* similarly touched on a search for meaning and belonging in contemporary male culture. Witnessed from outside the game, *Grand Theft Auto* appeared to pull men into a "toxic gospel," even a theocracy, over the years, as more and more bought into a violent online game world (or digital club).[70] However, inside the game, players discovered a variety of ways to belong in Los Santos, not all of them violent. In fact, the digital landscape often invited discovery and reflection just as much as simple acts of aggression. *GTA V* provided a range of unexplained mysteries and scenarios to explore and solve. Architect Frank Lloyd Wright wrote, "Tip the world over on its side and everything loose will land in Los Angeles."[71] The same went for Los Santos. Visiting Rockstar's creation was reminiscent of Steward Edward White's description of early Los Angeles as a realm of "electric signs," "crazy cults," and "fanatics."[72] Wandering the outer reaches of Los Santos felt akin to watching a 1990s *X-Files* episode, a real-but-unreal landscape of alien artifacts, nuclear waste, Big Foot hunting, military bases, and government secrecy. Hiding out in a compound in the Chiliad Mountain State Wilderness, the Altruist Cult, consisting mostly of luddite baby boomers, blamed other generations for the ills of the planet and likely ate human flesh. For some players, searching for meaning and myth became a key part of the game script. As one veteran player recounted, "I've beaten the game twice, and maxed out my stats, so myth-hunting is the only thing left to do."[73] In Thomas Pynchon's *The Crying of Lot 49* (1966), Oedipa Maas gazed down on a California suburb to discover, as put by Casey Shoop, a "circuit full of secret meaning."[74] Los Santos offered a similar realm of hidden meanings, positioned through a male gaze. Like Jameson, Chandler, Davis, and Pynchon, the player assumed the role of cartographer of Los Angeles, looking for its essence and hoping to find his own existence.

In *Fight Club*, the protagonist was a white-collar, white American with sleep and identity problems. In the *GTA* series, playable characters varied in their ethnicity and background, the common denominator being their criminal mindset. While most commentators accepted the white gaze provided in *GTA*, a number of critics voiced problems with the authenticity of the black experience offered. In the earlier game *GTA: San Andreas*, playing as Carl Johnson (known as CJ), an African American man residing in the South-Central district, represented on the surface a valuable opportunity for gamers to explore black masculinity. However, far from providing a realistic depiction of African American struggle, Rockstar instead presented a romanticized view of gang culture and the opportunities it afforded CJ. "CJ experiences hyper-mobility, exploring a world where he can not only become ultra-rich, ultra-fast, but where nothing is ever denied," noted Ed Smith, pointing to a "just take it" attitude in defiance of South-Central's realities.[75] Denying the problems of systemic racism, *GTA* goaded CJ's character with the simple line, "White people can do it, so why can't you?"

In *GTA V*, players could act out a black-and-white buddy game by swapping between Franklin and Michael. While Franklin emerged as the most likable of the characters, he depended on Michael for his success. Much like Rockstar's depiction of "stereotypical African American urban life: police corruption, gang warfare, drugs, and rap music" in *GTA San Andreas*, Franklin lived in a caricature world.[76] David Leonard calls Rockstar's approach one of "ghettocentric imagination."[77] By idealizing rap and gang culture, but rooting it in crime, the *GTA* series both celebrated and demonized black America. Scholars highlighted the deeper problems of playing "race" in video games—suburban whites playing ghetto blacks and taking on a temporary "black skin." Lisa Nakamura labelled the playing of racial stereotypes as a problematic exercise in "identity tourism," while Paul Barrett suggested that players assumed the black body as merely a style choice, a "commodified aesthetic," that lacked any depth or legitimacy.[78] As bell hooks noted in the 1990s, across popular culture "the desire to be 'down' has promoted a conservative appropriation of specific aspects of underclass black life."[79] Other video games, such as *We Are Chicago* (2016) by Culture Shock Games, provided far more realistic explorations of black masculinity. Based on extensive interviews with South Side residents, *We Are Chicago* represented a valuable attempt at gaming black culture. The player engaged with life choices over gang membership versus education, with the hope of a path toward a better life, as opposed to *GTA*'s pre-coded path of criminal activity.

In *Fight Club*, women acted chiefly as the fantasy of male eyes, the reader witnessing all events through the prism of the anonymous male protagonist. In the male fantasy of *GTA*, women were again seen through the eyes of male characters. Female characters remained paper thin, or, as one female journalist contended, "Women are *GTA*'s wallpaper."[80] Objectified as sex objects or nagging wives, women added little to the story. In "Visual Pleasure and Narrative Cinema," Laura Mulvey coined the term "male gaze" in reference to Hollywood cinema, noting how "in their traditional exhibitionist role women are simultaneously looked at and displayed, with their appearance coded for strong visual and erotic impact."[81] For Diego Gonzalez, *GTA* presented a clear example of the male gaze in video games.[82] In *GTA V*, Michael's wife, Amanda, once worked as a stripper, cheats on him with a tennis player, and hollers at Michael without refrain. Negative stereotypes abounded. The game provided no deconstruction of female characters, only satirical denigration. Rockstar defended Los Santos as essentially a male space. The *Guardian* situated *GTA* as "essentially an interactive gangster movie, and the genre has a long history of investigating straight male machismo at the expense of all other perspectives."[83] In 2013, Carolyn Petit reviewed *GTA V* for *GameSpot*. While rewarding the title with a score of nine out of a possible ten points, she offered a pertinent criticism that no decent women resided in Los Santos. Noting the paucity of positive female characters, with most there to be insulted or laughed at, Petit discovered, "an unnecessary strain of misogynistic nastiness running through [the game]."[84] A vitriolic backlash followed. An angry male fraternity defended the *GTA* world as sacred territory. *Grand Theft Auto* existed as a male fantasy and, as such, could not be criticized nor condemned.[85]

A City on the Edge

In the Joel Schumacher movie *Falling Down* (1993), Michael Douglas played embittered William Foster, a white-collar, white American having the worst of bad days in Los Angeles. Having lost his job as well as his marriage, Foster finds himself caught in downtown traffic and desperate for escape. He exits his car and decides to do the unacceptable— walk in Los Angeles. The film follows his journey across the city, an urban dystopia marked by gridlock and graffiti, ganglands and hatred, polluted skies and buzzing flies. As he negotiates his way through the urban decay, Foster finds himself equally unwelcome in both public and private spaces: gangs threaten to kill him in a dilapidated park-scape, while rich golfers threaten to evict him from their pristine fairways. Foster witnesses a world bereft of community or values, a world without social capital. The only decent citizen is a cop, planning on leaving the city upon his imminent retirement, who is tasked with finding Foster. Foster's descent into the bowels of Los Angeles is marked by his concomitant acquisition of weaponry. An ex-defense worker, Foster starts the movie with only his briefcase, but "upgrades" to a baseball bat, a knife, a bag of guns, and ultimately a rocket launcher. His weapon upgrades have a familiar gamic element to them; the continual swapping is a standard indicator of progression in a game. Filmed at the same time as the trial of four police officers who beat Rodney King and the ensuing riots, *Falling Down* imparted a sense of Los Angeles as a city on the edge.[86]

The *Grand Theft Auto* series similarly peered into the urban abyss. Rockstar depicted Los Angeles, in the guise of Los Santos, as a city marked by crime, looting, and automobile theft. The games helped reframe the mythology of Los Angeles around destructive spectacle. *GTA* was the latest in a series of novels, films, and other fiction exploring the decay of Los Angeles.[87]

Like *Falling Down*, Rockstar considered the idea of a city broken by racial hatred and ethnic rivalry. *GTA: San Andreas* explored the competition among the Bloods, Crips, and Hispanic gangs in 1990s Los Angeles. The climax of the game involved a recreation of the 1992 riots. In the gamic San Andreas, the riot begins after corrupt LAPD officer Tenpenny is arrested for murder and drug dealing, but the district attorney drops the charges against him. The radio announces the "surprise decision," with a disc jockey noting that "cops always get off easy" and "Los Angeles will burn tonight." Gathered at a local gang house, CJ hears the news. A hippy friend says, "Power systems corrupt everyone," while another opines, "People real mad . . . as if the ghetto ain't wrecked enough." CJ then embarks on a mission to wade through the riot zone and "lock down" his "hood." He drives through downtown: gangs shooting, cars on fire, rifles up for the taking. A police car explodes. The radio announces that "the whole city is a war zone." The fall of Los Angeles happens around the player, as CJ sees and participates in the real-time collapse. In the final mission of the game, CJ steals a locally deployed SWAT vehicle caught in the riots so that he can kill, in vigilante style, arch-villain Smoke. Tenpenny is killed on the street, while an African American character jibes, "See you around, officer." Rockstar's simulation of the riots was flawed. With no mention of historic racial divides, poor quality housing, economic deprivation, or the 1965 Watts riots, *GTA* provided little background to the troubles. Paul Barrett critiqued the unfolding "urban nightmare" in *GTA: San Andreas* as unexplained and de-historicized. With Tenpenny cast as a black American, Rockstar misrepresented the riots as a gang squabble, rather than a product of enduring racial conflict. The *GTA* riots symbolized "a black problem for black people."[88] Ed Smith similarly contended, "Rockstar excludes from its version of the LA riots any examination of historical racial tension"; thus, "one of the most politically and socially significant moments in 20th-century America . . . is made apolitical and asocial."[89] While Rockstar highlighted police tensions with the African American community, it also transformed rioting into action sequence and visual spectacle. Los Angeles seemed on the edge of collapse, but without a full explanation.

The sense of Los Angeles as on the edge could be seen in other Rockstar titles. The detective title *L.A. Noire* (2011) paid homage to Los Angeles of the 1940s, corrupt, dark, and decried by working men as "the scabbiest town on earth."[90] The title explored a world of crime, vice, and degeneracy. Rather than a city of angels, Los Angeles existed purely as a city of criminals. Using street maps and city plans, Rockstar slavishly recreated downtown Los Angeles. Relying heavily on period photography by Robert Spence, programmers accurately simulated an eight-mile-long stretch of historical Los Angeles.[91]

L.A. Noire also paid homage to the classic noir thriller. Crime writers had immortalized Los Angeles as a crime landscape marked by murder and revenge. The title gamified the noirish city imagined by Raymond Chandler and James Ellroy. As hard-boiled LAPD detective Cole Phelps, the player set about solving a series of crimes, some based on real cases, including the Black Dahlia. The game behaved as a classic detective novel. Players navigated a virtual Los Angeles as a city of purely criminal activity, constantly looking for the next incident. In a nod to Chandler, the in-game movie theater played *The Big Sleep* while players won accolades that included "The Simple Art of Murder." Like Marlowe driving across town, players took to the road to investigate murder after murder. The darkness of the game provided a constant reminder that things were not right, that murder marked the city as on the edge of a precipice.

In 2013, *GTA V* provided the most detailed installment of the "city on the edge" narrative. The opening sequence involved the player in an armed robbery of a bank and a subsequent police chase. In *L.A. Noire*, the player stopped crime; in *GTA V*, the player committed it. In stark contrast to the crime fiction of Edgar Allan Poe or Arthur Conan Doyle, in which "crime is an aberration," *GTA* embedded crime in the fabric of Los Santos, with violence remaining "a pervasive feature of the urban landscape."[92] In *GTA V*, crime seemed everywhere, with even "the city itself is corrupt." Compared to Mike Davis's mapping of Los Angeles as an "ecology of fear," Rockstar presented the city as an ecology of deviance. Davis described Los Angeles in the 1990s as being in a perpetual state of fear of crime, resulting in a closed and repressed society, with gated communities determined to keep the nasty public out and the affluent private in. "Anyone who has tried to take a stroll at dusk through a strange neighborhood patrolled by armed security guards and sign-posted with death threats quickly realizes how merely notional, if not utterly obsolete, is the old idea of the 'freedom of the city,' " wrote Davis.[93] In *GTA*, players took on the role of the enemy feared by the gated community, with the "freedom of the city" to roam and destroy.

Like *Falling Down*, *GTA V* depicted an urban environment caught in social, environmental, and cultural decay. Gangs operated outside the law, their graffiti tags marking their growing territory. Los Santos citizens seemed lost, unfriendly, or downright dangerous. A radio announcer prefaced a song, "This is dedicated to the lost souls of Los Santos." Players navigated an Autogeddon landscape filled with smog, gasoline, and waste. One task for the player involved collecting thirty nuclear waste barrels spread across Los Santos. The waste revealed the growing contamination of the region. Highlighted in plastic surgery commercials ("Why not try an extra nose or three breasts?") and celebrity gossip on Weazel news (MC Clip explaining how "I'm going to take humanity to new places" on his album, *The New Messiah*), Rockstar documented the widespread decay of city culture. Los Angeles seemed trapped in low-quality reality television. City life perpetually disappointed. Having moved to Los Angeles for "the weather" and "that magic. You see it in the movies," Michael found the "miserable" reality one of "fairy tales spun by people too afraid to look life in the eye."

At the end of *Falling Down*, suddenly realizing he has become the bad guy, Foster commits suicide. At the end of *GTA V*, the player faces a choice of which main character dies.[94] Both film and game carried a sense of cynicism and declension, providing an anti-Hollywood happy ending and a mocking of the California Dream in tune with Los Angeles's darker side. Los Santos seemed a place on the edge, where people always fell down.

Los Santos / Lost America

Send us your brightest, your smartest, your most intelligent,
Yearning to breathe free and submit to our authority,
Watch us trick them into wiping rich people's asses,
While we convince them it's a land of opportunity.

INSCRIPTION, STATUE OF HAPPINESS, GTA IV

In 1995, California visual artist Sandow Birk painted a series of evocative landscape pictures entitled "The Rise and Fall of Los Angeles." Birk looked to Hudson School painter Thomas Cole's "The Course of Empire" for inspiration. Cole's work was a magnificent set of canvases produced in the 1830s that depicted and satirized America's epic rise to world dominance and concomitant fall. In an age in which the pastoral and agrarian idyll still flourished, Cole painted the rise and fall of an imaginary American city. The impermanence of the built landscape contrasted with the longevity of nature. The wild ultimately subsumed the American civilized.[95]

While Birk's first canvas depicted a "Savage State" of dinosaurs, and the second, a pastoral scene, his third, the "Consummation of Empire," captured the distinctiveness of modern Los Angeles in the form of skyscrapers, freeways, and billboards. Birk evoked the specter of consumer excess in a landscape dominated by banal advertising. Only the graffiti tags of a "lower class," gathered in the darker spaces of the painting, beneath the underpass, suggested signs of originality as well as unease. Beck's fourth image, "Destruction of Empire," resembled a scene from a Hollywood disaster movie. As if a visual embodiment of Mike Davis' *Ecology of Fear*, everything was coming apart, with a woman falling to her death and a juggernaut hanging over a broken freeway.[96] Birk's final image, "Desolation," depicted the Los Angeles hundreds of years into the future, with greenery having enveloped the remnants of long-wrecked freeways and skyscrapers. This Los Angeles resembled a ruined Pompeii or a fallen Roman Empire.

Birk employed a familiar technique of using Los Angeles as a dark mirror to reveal the broader fate of the United States. Early city boosters had presented Los Angeles as an "Earthly Paradise" where anything was possible, where the American Dream could finally be realized—a common tactic for advertising American frontier towns to the East. Skeptical of exaggerated claims, Louis Adamic wrote of "high priests of the Chamber of Commerce whose religion is Climate and Profits."[97] Behind the billboards of betterment lurked lingering embitterment. Right from the start, the "Dream" invited nightmares. Writers visiting Los Angeles for the first time in the 1920s found themselves drafting cheap scripts for B-movies: "For these newcomers, L.A. was exotic territory, full of outsized ambitions and outsized fantasies, a funhouse mirror offering greatly exaggerated reflections of America itself."[98] However, the American Dream existed only on the Hollywood screen, and the writers found themselves destitute. Boosterism gave way to skepticism thanks to social critics such as Sinclair Lewis. Rather than a mirror of America the great, Los Angeles showed only the worst. In his diary for 1926, Adamic described a realm of growing chaos, declaring, "Los Angeles is America. A jungle . . . It is a bad place."[99] On visiting the town in the same year, journalist H. L. Mencken declared, "There were more morons in Los Angeles than in any other place on earth."[100] In *The Day of the Locust* (1939), Nathanael West depicted Los Angeles as a realm of simmering violence. West provided a glimpse of the larger unfolding of the LA project. Rather than Los Angeles as "future America," a new city on the hill, West cast Los Angeles as a nation unravelling. As Lynn Mie Itagaki imparted, "California is the land of extremes, home to the grandest American dreams and the harbinger of national or global collapse."[101] By the 1970s, Los Angeles seemed firmly ensconced on the frontier of a dark America. Carolyn See declared that "the West Coast is the end of the road for the American Dream," while Mike Davis wrote of literature that forwarded Los Angeles as a "nightmare at the terminus of American history."[102]

Rockstar used the same dark mirror to reveal similar conclusions about the nation's fate. The sense of dystopian spectacle that characterized Birk's paintings infused Rockstar's digital canvas. Rockstar critiqued the American Dream through the dystopic lens of Los Santos. *GTA V* depicted the rise and fall of America through a landscape dominated by repetitive images of consumer advertising and excess and marked by uncontrollable riots and mass shootings. Like a demented joyrider, the player drove through a landscape of violence and decay. Los Santos was depicted as a world of wealth, vice, and corruption slipping into a final stage of barbarism. Rockstar provided a game based around American decline. As Houser simply put it, *GTA V* explored "the endpoint of the American Dream."[103]

Rockstar accounted for the demise of America in a number of ways. It pointed to problems of capital, race, and consumer culture, and presented access to the American Dream as a critical factor in its failure. In *GTA IV*, the Statue of Happiness (a Rockstar version of Lady Liberty), featured a chained, beating heart inside, with a cynical rewrite of Emily Lazarus's poem "New Colossus" as inscription, underlining an unwelcoming nation. Both East European immigrant Niko in *GTA IV*, set in New York, and African American Franklin in *GTA V* failed to achieve their goals, while Michael, also in *GTA V*, explained, "I'm living the dream, pal, and that dream is fucked." With its lead characters continually disappointed, Rockstar's *Grand Theft Auto*, like F. Scott Fitzgerald's *The Great Gatsby* (1925) before it, tackled the elusive American Dream of riches and happiness. Seekers of the American Dream, rich and poor alike, found their hopes crushed in the game world. Like readers of John Fante's *Ask the Dust* (1939), the audience of *GTA* faced the task of "dismantling the glittering promises of the dream city and finding unsettling realities underneath."[104]

Rockstar highlighted the contribution of modern consumption to American collapse. Like Nathanael West satirizing consumer life in the 1930s, Rockstar explored "the kindling and betrayal of consumer dreams" in the 2000s, with disappointed consumers everywhere.[105] Drawing on the reputation of California for "excessive waste and superficiality—expensive and inefficient cars, overwatered lawns, McMansions, and far-flung bedroom communities," Rockstar highlighted flagrant consumer excess.[106] From Taco Bomb ("fed students, poor people and drunks since the 1970s") to Logger beer ("A classic American tasteless, watery beer; made by Rednecks, for Rednecks. Now Chinese owned"), every item seemed tainted. In its humorous parodies, *GTA* bore similarities to *The Simpsons* and *South Park*. Houser blamed the US media and mass advertising for heightening the problems of consumer society, noting that all Rockstar games "are set in a world that is a satire of American media culture."[107] A play on Facebook, *GTA*'s LifeInvader featured the motto "Invasion never felt so good," with one mission based on infiltrating the organization and killing its owner. All kinds of entertainment, including video games, served as objects of satire. Michael moaned to his therapist about his son, who "sits on his ass all day, smoking dope and jerking off while he plays that fucking game."

Rockstar depicted an America overwhelmed by corruption and the excesses of capitalism, a world in which FBI and CIA agents fashioned fake terrorist threats to justify their work while selling drugs on the side. The *GTA* series repeatedly highlighted the "uniformly corruptive" influence of the dollar through a world populated by elongated limousines, gangster bling, and misleading advertising.[108] Even justice could be bought, as seen in the advertisement for "Slaughter, Slaughter, Slaughter: Prominent US Law Firm, helping guilty people get away with it for over 30 years." This gamic America was also a land of increasingly angry and dejected people. Standing across the road from LS Customs, an automobile respray garage in the desert section of Los Santos, a man wore an Uncle Sam outfit and shouted, like a doomsayer, about the perils of immigration, while pointing to the end of the world. Michael shouted at the Pizza Store, "Fuck your pizza. Fuck everything it stands for." Cast as a "bubbling cesspit of celebrity fixation, political apathy and morose self-obsession," Rockstar's America appeared fundamentally broken.[109]

In 2005, NBC's Kristin Kalning commented, "The Grand Theft Auto games are shorthand for everything wrong with America. Violence. Prostitution. Drug smuggling. Gun running. Political Corruption. Racial tension."[110] Highlighting Rockstar's continual satire of US life, *Salon. com* headlined an article "GTA Hates America."[111] Triggered by the discovery of secret in-game sex scenes (the "Hot Coffee" scandal) in *GTA: San Andreas*, US Senators Joseph Lieberman and Hilary Clinton led a Federal Trade Commission investigation into Rockstar, resulting in a $50 million fine for distributor Take Two. Clinton cited *GTA* as a threat to American values and suggested, "We need to treat violent video games the way we treat tobacco, alcohol, and pornography."[112] Concerned citizens feared the corruptive influence of violent gaming. Arguably, the magic circle of gaming no longer held true. In Baton Rouge, Louisiana, a young man charged with hit-and-run driving told a police officer that he "wanted to see what it was to be a Grand Theft Auto individual, what it felt like."[113] The boundaries of gaming appeared breached, as people drove in Los Angeles while thinking of Los Santos.

The game's defenders highlighted its worth as a satirical mirror, casting a gamic lens back on the society that produced it. Dennis Redmond applauded Rockstar's "willingness to shine a spotlight on the dank underbelly of the US empire," while Joris Dormans argued, "If the game is subversive, it is so not because it teaches youngsters to be criminal, but because it teaches them to appreciate their society critically."[114] Rockstar outlined in the most colorful of ways the challenges of American society. Rather than a fantasy-based take, the game offered satirical realism. Ed Smith, writing in *Kill Screen*, noted that "[the game's] nightmarish violence isn't even that fantastical anymore with a mass shooting on the news every other week."[115] Kelly McDonald, writing for the *Guardian*, remarked, "Its America is hyper-saturated, hyper-satirical and hyper-violent, but the game's preoccupations—the corrupting influence of capitalism, sensationalist media and the impossibility of living a good life in a world with very little goodness in it—are very much grounded in the real world."[116]

Rockstar presented a broken America as a rollercoaster thrill ride, and most gamers loved it. Full of contradictions and paradoxes, it rarely conformed to one underlying concept. While the company poked fun at American values, most players still set out to increase cash and collect cars, reinforcing the same capitalist system under attack. In 1926, British writer Aldous Huxley, staying in Santa Monica, wrote "Los Angeles. A Rhapsody." He described a world of studios, fast-moving images and fast cars, religious extremes, fake women, and constant movement, "where thought is barred" and "conversation is unknown."[117] Huxley defined Los Angeles as a flawed project, but at its core, a "good time." He called it the "City of Dreadful Joy." *Grand Theft Auto* seemed not much different.

7

Second Life, Second America

IN THE FIRST DECADE of the twenty-first century, hundreds of thousands of Americans logged onto an online world created by Linden Lab, a new technology company based in San Francisco. The virtual citizens wandered neon-signed shopping malls looking for clothes. They chatted, danced, and got married. They had affairs. They bought sex at street corners. A few went to church. Some built their own properties, others rented. On weekends, they visited the Lincoln Memorial and the White House, or trekked across Yosemite and took flights over the Grand Canyon. Some even travelled to the Apollo II moon landing site. They inhabited an online platform called *Second Life*. It offered an alternate world, a Second America.

Commercially released on the PC platform in June 2003, *Second Life* offered an expansive virtual territory to explore. Linden furnished the software tools for people to fashion online lives in a vast, user-generated realm. "Residents" could design their own avatars and construct their own homes, find romance and adventure, and, within a few months of release, even take online jobs to earn Linden dollars. *Second Life* welcomed all and sundry to a virtual existence. A huge exodus of people left behind evening television, exercise, and hobbies to enter the virtual domain. Soon, real-world companies opened stores in *Second Life* streets, universities set up campuses, news journalists covered in-game stories, and US politicians campaigned there.[1] *Second Life* emerged as a cultural phenomenon. It seemed everyone wanted a virtual existence.

The Game of (Second) Life

From the outset, Linden Lab sought to differentiate its product from the traditional market of video games and electronic entertainment. The fledgling San Francisco company presented *Second Life* more as "a new experience" than as a game. CEO and founder Philip Rosedale said, "We don't see this as a game. We see it as a platform that is, in many ways, better than the real world."[2] He adamantly refuted the idea that the title could be pigeonholed as a simple computer program. *Second Life* signified a "concept" and a lifestyle choice. Linden situated its program as a distant cousin of the modern video game. The creation of a world ostensibly without targets, high scores, or game rules challenged traditional notions of the ludic. The formation of a world without a guiding story or an overarching direction undermined conventions of narrative. Linden envisioned *Second Life* as a piece of software far more expansive in scope than traditional computer entertainment and capable of ushering in a new realm of online interactivity.

Second Life was nonetheless a game, of sorts. Rosedale envisaged the title as providing a "huge playspace" for its members, somewhere that people could congregate and have fun. First-generation avatars wandered a virtual world rich in ludic prospects.[3] Initially, play took the form of combined construction projects using PRIMs (primitive building blocks). As software tools evolved, residents created various in-world games, from simple "coin capture" to hunting and sports simulations. Gradually, *Second Life* catered to all kinds of ludic impulses. Far from the obvious mechanics of run and jump in a Nintendo *Mario* game, Linden's software arguably promoted a more generous and inclusive sense of play than any video game title. Play could mean anything from sexual foreplay to choreographed disco dancing to skating in a virtual Winter Wonderland at Christmas. Defying traditional rules and gamic structures, *Second Life* promoted a form of unstructured play true to the concept of "paidia" first mapped out by French sociologist Roger Caillois in the 1950s.[4] In a sense, gamers were left to define both the breadth and limits of computer entertainment.

Second Life also had its origins in classic video game design. The user interface, avatar control system, and three-dimensional appearance all mimicked conventional gaming templates. In the prototype version *Linden World* (2001), players controlled third-person avatars and explored blocky three-dimensional environments. Populated by gun- and grenade-wielding, jet-packing avatars (called primitars), *Linden World* looked conspicuously like such third-person games as DMA Design's *Body Harvest* (1998) and not that different from first-person shooters *Doom* and *Duke Nukem 3D* (1996). Destructible environments and powerful weaponry suggested a game more about shooting aliens than shooting the breeze. The new world of immersive, user-generated play seemed, at least at first, firmly rooted in the old world of gaming.[5]

Linden's plans for a user-created world also referenced another gaming template: the multi-user dungeon (or MUD). In the 1970s, the dice-based role-playing game *Dungeons & Dragons* (D&D) catapulted the MUD into suburban basements and into cult status. In the United States, a whole subculture of D&D players emerged. With video games such as *Adventure* (1976), programmers transferred the MUD template to computers. In 1987, LucasArts released a pioneering networkable title, *Habitat*, for the Commodore 64 home computer. Players assumed control of cartoon-like characters and communicated with each other via speech bubbles. Hugely ambitious in scale, the game world included cities surrounded by wilderness. *Habitat* provided its players with a token economy, housing, and a large degree of freedom. The inventiveness (and transgression) of early residents led the designers of the game to impose rules on the fledgling society. Rampant gun culture and an unwelcome outbreak of violent crime, including the killing of avatars, led to tight restrictions on in-game firearms, with shooting allowed only outside of towns.[6]

In the late 1990s, massively multiplayer online role-playing games (MMORPGs) such as *Ultima Online* (1997) and *Everquest* (1999) superseded *Habitat*. Appealing to fantasy players and resource managers alike, titles such as *World of Warcraft* (2004) generated vast user bases. By 2007, an estimated nine million users inhabited *World of Warcraft*. Linden Lab backed a system of online play similar to these titles. However, the totally unscripted nature of *Second Life* set it apart from most MMORPGs. Linden offered players far more freedom than *Habitat* or *Warcraft*. Granting individuals the tools to create their own content, Linden shifted the dynamic of creative imagination onto the gamers.

Rather than inhabit a company-defined fantasy, players could concoct their own fantasies. The ability to create an 'own world' in *Second Life* led to vast amounts of home coding and modifying (or modding).

This concept of "making your own life" allied *Second Life* with the simulation genre of gaming, in particular the highly successful franchise *The Sims* (2000+). Created by Will Wright, *The Sims* allowed players to run their own virtual households and families, with the player managing Sims in their diets, hygiene, work lives, and time spent moving furniture. *The Sims* offered a carefully choreographed simulation of American life based on building, controlling, and consuming. It provided an entertaining simulation of the American everyday. In December 2002, *The Sims* went online. Initial excitement gave way to disappointment over the restricted nature of interaction, indicating a notable gap in the market, soon to be exploited by Linden.[7]

Twenty-first-century gamers welcomed the shift toward more comprehensive simulation. While many video games focused on specific activities or scenarios (from playing tennis to fighting terrorists), the potential for a more holistic simulation of life materialized thanks to advances in computer technology. While in the past, specific elements of nature or history had been simulated, now life itself could become digital play. The growing complexity of the avatar reflected this trend. From elemental line drawings, the avatar had come to embody something deeply personal and identifiable to players. From the first "stick men" in *Gun Fight*, the video game industry had evolved player representation to encompass individual style, gender, and body tastes. *Second Life* granted players full creative control over their avatars, with the ability to change hair and skin color, facial and bodily expressions, and sartorial complement. *Second Life* represented the next step in avatar evolution. Such increasing avatar complexity drew the player closer to the game. As philosopher Guy Debord once described life as "an integral thrilling game," *Second Life* forwarded the concept of "living" in a game.[8]

In its open world design and capacity for user-generated content, *Second Life* reflected a potential next stage in computer entertainment. It was also part of a broader cultural exploration of virtual possibilities. Both the rhetoric and realities of a new digital age informed the game space. In 1992, Neal Stephenson coined the term "Metaverse" in his science fiction novel *Snow Crash*, a story of high-tech anarcho-capitalism set in a dissolved United States. Metaverse described the next evolutionary stage of the Internet, in which user-controlled avatars traversed a single 65,536-kilometer street, with the virtual real estate along it run by the Global Multimedia Protocol Group. The concept of a shared three-dimensional virtual space inhabited and molded by avatars, but facilitated by an outside corporation, matched the basic Linden model of *Second Life*.

Other science fiction works probed distant stages of virtual reality and online existence, exploring not just a second life, but what a third or fourth incarnation of digital-based "living" might encompass. Brett Leonard's 1992 film adaptation of Stephen King's short story *The Lawn-mower Man* highlighted the potential for virtual technology to liberate people from their human bodies and their disabilities. The Wachow-skis' *The Matrix* (1999) explored the distant possibilities of computer-generated alternate worlds and the point at which humans swap from material existence to simulated reality. In *The Matrix*, sentient machines enslave humans, trapping them in a dream state. Rebellion guru Morpheus asks the protagonist, Neo (played by Keanu Reeves), "Have you ever had a dream, Neo, that you were so sure was real . . . How would you know the difference between the dream world and the real world?"

In the real world, game companies experimented with elemental virtual reality technology, but with little mainstream success. The Nintendo Virtual Boy launched in 1996, promising "a 3-D game for a 3-D world" with dramatic adventures in a widely promoted "third dimension."[9] Users suffered intense headaches from the associated motion sickness. Instead of complex three-dimensional fantasies, the virtual world took the form of more practical online interaction, including email, messenger services, and chatrooms. Rather than a quantum jump to a new realm, "real" society plugged into the virtual society in a more gradual and integrated way. Millions of Americans logged on to the Internet to join social media platforms such as Myspace and Facebook, as well as chatrooms such as AOL Instant Messenger. The Internet connected people in new ways. *Second Life* captured the aspirations of the new Net Generation by bringing together a range of communication tools (instant messaging, chatrooms, etc.) in a cohesive, realized space. It tapped new technologies and habits. Rather than spearheading a revolution, *Second Life* was part of a broader cultural and technological transition toward a digital society. The growing popularity of gaming, virtual gatherings, and social media highlighted an American community clearly spending far more time online, but the notion of a total and instant break from real world society was a chimera. Instead of trading one realm for the other, practices in the real world and the virtual world merged.

The Great Migration

In March 2002, the demonstration, or "alpha," version of Linden World opened to select guests. The first official resident, a Californian mother and web-designer, Steller Sunshine, stayed there overnight, while she fashioned a first settlement. The digital sun rose to a lone home on the horizon, consisting of a cabin and a beanstalk. Skyward thinking was tethered to an old frontier base. The brave new world of *Second Life* seemed fantastical and otherworldly, but also strangely familiar.[10]

While organic and freeform, visions of a new community in *Second Life* resembled versions of the New World firmly rooted in the American zeitgeist. Sunshine's beanstalk referenced fantasy and folk story, but her cabin spoke to classic American pioneering. Digital fantasies stemmed from material design. Concepts of a virtual New World rested in traditional American history and folklore. Popular references to Columbus framed the expansion into digital worlds as a repeat exercise in colonial navigation, discovery, and conquest. Like Spanish explorers landing their galleons on the Eastern coast of North America, the first Second Lifers logged on to discover a land of abundance ripe for cultivation.

Linden expected a gradual migration of real-world users to its virtual domain, with settlements, commerce, and societies all forming in the blank space of Da Boom (a new land named after the Big Bang). The *Second Life* experiment resembled an exercise in nation-building—a virtual re-enactment of the processes that fashioned real-life countries. Given that *Second Life* heralded from San Francisco and Silicon Valley, with early participants often American in nationality, and US news media covering the story, parallels with the development of the United States seemed unavoidable. American technologies, values, and commerce, as well as a frontier mentality, shaped the emergent game space. Digital pioneers could hardly delete their own background "programming" on entry. Rather than being a space beyond countries, as imagined by cyber-libertarian John Perry Barlow, digital realms quickly took on real-world rhetoric, allusion, and meaning.[11] At times, *Second Life* reflected a distinctly American cultural hegemony and trajectory.

The migration attracted a wide range of participants. A mix of social gamers, role players, innovators, technology enthusiasts, programmers, Internet users, and chatroom fans headed for the New World. The broad appeal of *Second Life* rested on notions of escape, release, and novelty, as well as a craving for a different kind of life. *Second Life* promised a fresh start for all, regardless of background.

The migration also sported a strong frontier rhetoric. In his "Notes from the New World," journalist Wagner James Au described the migratory process as akin to nineteenth-century westward migration.[12] For Au, Linden land parcels resembled the Homestead Act of 1862. Choosing a Hunter S. Thompson–type avatar, Au likened himself to a "reporter of a frontier-town newspaper" who had a duty to document the arrivals. By contrast, author and entrepreneur Mark Stephen Meadows paralleled the rise of *Second Life* with developments on the twentieth-century urban frontier, specifically the rise of Los Angeles in the 1920s.[13] For Meadows, the same impulses of boosterism and land speculation, as well as the lures of a perfect climate and media interest, drove both developmental projects. A similar sort of "hype" to that which coalesced around Los Angeles attracted a mass of immigrants to *Second Life*, "each hoping to strike it rich, become a star, find a friend, or simply explore new possibilities." For Meadows, *Second Life* and Los Angeles shared similar dynamics; both presented "a land of profitable automation, easy luxury, inexpensive fun, and independence."

Part of the new technological age, *Second Life* seemed caught in a broader romantic conception of cyberspace as America's next frontier. Journalist Jeffrey Cooper talked of "a vast new territory" being created by digital technology in the 2000s; "in doing so, they have reopened 'the frontier.' "[14] Referring to the growth in online worlds and virtual reality, Cooper declared, "The new cry should be not, 'Go West!' but 'Go Cyber!' " Technology promised a new realm for American freedom. Howard Rheingold's *The Virtual Community: Homesteading on the Electronic Frontier* (1993) described his embracement of WELL (Whole Earth 'Lectronic Link) in the early 1990s as "participating in the self-design of a new kind of culture," a "speeded-up social evolution" and an "authentic community."[15] Meanwhile, *Motherboard* technology magazine described Linden's project as "a new frontier for exhibitions, installations, and performances."[16] *Second Life* was part of a much more expansive digital revolution.

On the first anniversary of *Second Life*, residents gathered for a gala parade organized around the theme of "digital pioneers." On October 18, 2006, *Second Life* welcomed its one millionth pioneer. Within three years, the population of the online world had gone from less than one hundred to rival that of a medium-size, real-world city. Commentators interpreted the great migration as part of something even bigger: a fundamental shift in society whereby people spent significant time in virtual worlds. Writing in May 2007, Gartner Inc., a technology analysis company, predicted four out of five Internet users "would have a second life" within just three years.[17] The majority of Americans were expected to make the virtual migration.

Remaking the American Landscape

Alongside the individual projects constructed in the alpha version, the Linden team designed and constructed Linden Town. An experimental settlement to gauge how building tools performed, Linden Town represented the first organized construction project in the online space. The collection of buildings resembled a small American conurbation. In homage to the real-world location of Linden Lab, the team recycled San Francisco street names for the fledgling hamlet. Linden featured a town square, fountains, several roads, and a city hall. Outside the city hall, they crafted a huge statue called simply "The Man."

Linden Town resembled a twenty-first-century Jamestown. In May 1607, ships on behalf of the Virginia Company of London landed in Chesapeake Bay on the Eastern seaboard of America. The settlers, mostly with aristocratic and entrepreneurial backgrounds, came to the new land with hopes of trade. They envisaged the wild terrain as a landscape of opportunity. Some looked for escape from persecution, for a fresh start, others hoped to return to their old lives in Britain with mercantile riches. The corporate project of Linden Town marked the beginning of a similar pioneering venture. Through membership dues and taxation, digital pioneers entered the virtual territory under the bind of the Linden corporation. They hoped for escape from their real lives, as well as the discovery of fresh entrepreneurial opportunities. The architects of *Second Life*, including Rosedale himself, were the new "founding fathers" of a digital experiment. Images of *Second Life* as a land of opportunity recycled familiar American myths and motifs. America 2.0 resembled America 1.0 in its focus on escape, capital, and accumulation. Like the Captain John Smith statue erected at Jamestown, the founding father of the Linden "Man" projected good fortune across the settlement. Gradually, Linden Town dissolved as new welcome areas took its place, but the Man remained as a powerful symbol of origin and destiny. Overlooking a sea of virtual water, the imposing statue welcomed waves of new migrants, greeting them like the Statue of Liberty off Manhattan.

Free to build what they wanted, digital colonists assembled a panoply of wild and extravagant structures. The virtual land grab inspired both large- and small-scale projects. Science fiction fans constructed the city of Nexus Prime, a huge cyberpunk creation begun in March 2003. Groups banded together to generate artistic installations and large-scale projects. Often, chaotic and unplanned architectural forms dominated. The artistic freedom granted to thousands of new users translated into a maelstrom of styles. In the absence of any overriding urban plan, *Second Life* lacked regimen or order. By 2010, territory stretched to 1.8 billion square meters, a landmass larger than Houston, Texas, but with little obvious coherence.

Alongside highly imaginative and otherworldly edifices, *Second Life* denizens often chose to build structures inspired by the real world. Architectural choices reflected the longing of residents for a degree of realism throughout their fantasy. Gamers chose to construct abodes that mimicked real-world conurbations. They enacted a practice of "artificial realism" not that far removed from Disney's facsimile of Main Street in Anaheim's Disneyland. *Second Life* provided a second version of familiar sights and places.[18]

Such architectural familiarity reflected an unexpected behavior of residents. Despite occupying a virtual realm unimpeded by traditional laws or concepts, residents proved keen to replicate real-world practices. Most players chose human rather than animal avatars. Rather than fly to places (an early ability coded by Linden), most people walked. Many seemed reticent to embrace the breadth of freedoms provided by Linden. Just like outside the game, people acquired land, furnished houses, and even went to work. *Second Life* mirrored first life. The New World seemed not that dissimilar from the old world.

This search for the familiar often assumed American dimensions. Alongside unfolding techno-futurist and steampunk realms, new residents began shaping specifically US-inspired projects. American influence could be seen in early settlements, geography, and architecture.

In early 2003, resident George Busch operated a Route 66 diner in the fledgling Blue sim district of *Second Life*. Sinatra Cartier (real-world name: Rick Ellis) ran a Route 66 bar nearby. Joined by four other *Second Life* devotees, Busch and Cartier began planning a far more elaborate and expansive sim project. The team discussed the idea of a themed world dedicated to US values. Architect Tracey Kato, who joined *Second Life* in December 2002, forwarded the concept of a landscape celebrating "the best of the USA, from Miami Beach to L.A. and everything in between," a world with room for "anything and everything that helped make our country great (and maybe the not so great)." Kato's definition suggested, in essence, a landscape of American architectural symbols. The project would join together a range of individually authored monuments inspired by the real world, including recreations of Fenway Park, Washington Monument, the Empire State building, and the Chrysler building. The project would be called "Americana."[19]

Gradually, the contours of a virtual American territory took form. The team constructed the White House, Lincoln Memorial, and the Statue of Liberty. Historic landmarks marked the emergent cartography of *Second Life*. Familiar architecture transformed part of the blank canvas of *Second Life* into something instantly identifiable. Americana existed as a navigable collection of potent architectural icons. It combined the monumental, the utopic, and the tourist friendly. The group-owned sim functioned as a theme park dedicated to the theme of nationhood. In 2005, it became part of a broader World's Fair–style "American exhibition" held online.

Other American-themed projects followed, but none on such a dramatic scale. Elsewhere in *Second Life*, individual residents chose specific buildings, time periods, and places to emulate. *Second Life* Miami Beach existed as a beach, a few deck chairs, an outdoor club, and some shopping. Most simulations essentialized the American experience. Like Disney's facsimile of Main Street USA, designers took the popular essence of a place and rendered it in virtual form. *Second Life* offered a dream-like digital fantasy of a nation. For Au, *Second Life*, with its facsimiles and visual furnishings, amounted to an "Impression Society." Residents created impressionistic virtual shrines to the material landscape.[20]

The roster of virtual monuments included facsimiles of the great outdoors. In the late nineteenth century, the transcendental and romantic movement celebrated the American wilderness as the true home of national character. Landscapes such as Yosemite, Yellowstone, and Niagara Falls marked the distinctiveness of the American experience and served as geographic hallmarks of a new nation. As Alfred Runte first noted, commentators offered such places as America's "monumental scenery," a form of cultural capital to compete with the venerated cathedrals and castles of Europe.[21] Second Lifers engaged in a similarly patriotic act by crafting statements of American natural greatness in new electronic realms. Virtual nature took on a patriotic slant. Authors crafted a form of virtual monumentalism.

Created by Cara Caldi in 2011, Walden Pond in Botanica celebrated nineteenth-century Concord, Massachusetts, and its most famous literary habitué, Henry David Thoreau. A recreation of Walden in the 1840s, Caldi's rural idyll featured a range of impressive foliage encircling a restful lake. Perpetually in autumn, rust-colored leaves drifted to the ground, ducks swam on the cold water, and animals scavenged in preparation for the winter, including a wary black bear wandering the trail. Rather than frozen in time, Caldi's landscape felt alive and inviting. Visitors entered a simulation of rural Concord in the mid-nineteenth century. *Second Life* provided a portal through time.

Visitors to Walden Pond entered a distinctly Thoreauvian fantasy. A virtual "life in the woods" drawn directly from his prose, Caldi's Walden was rich in atmosphere and detail. Positioned just a few steps from the forest path stood Thoreau's wood cabin, complete with writer's desk and ink blotter, "fresh" eggs, and a portrait of the author above the mantelpiece. An axe hung from a stump of wood outside, waiting to be taken, as if a distracted Thoreau had wandered off into his own dreamscape and would return imminently. Strolling along the winding path and peeping inside the cabin, visitors entered the writer's imagination. The landscape worked as a shrine to America's most famous transcendentalist.

Caldi encouraged her audience to directly engage with the Walden experience. Rather than providing a simple walk in the woods, her version of Walden featured integral didactic aspects. Instructed on entry to search for notable quotes from the transcendental thinker etched on a series of large stones, visitors foraged in the wilderness not for food, but for wisdom. For those who discovered all eleven extracts from *Walden* scattered across the landscape, a reward lay at the journey's end. The successful player received either an animal friend or a copy of Thoreau's journal to take away with them. As Thoreau himself imparted, "Books are the treasured wealth of the world and the fit inheritance of generations and nations."[22] In the process, Caldi gamified *Walden*.[23] She transformed a nature experience into a collecting game with educational components. With the foliage of specific trees and animals for sale, marked by Linden price tags, visitors could even "buy" the Walden experience. The designer commodified the transcendental moment. Such abject commercialism undermined Thoreau's own plea to escape a material-bound existence.

The search for wilderness in the virtual realm of *Second Life* also included recreations of several US national parks. Yosemite existed first as a sim in 2012 designed by Sominel Edelman; then, in 2016, Maganda Arts led by Jadyn Firehawk opened Yosemite National Park online. Rather than attempt a geographically accurate rendition of the wilderness territory, Firehawk reduced the Yosemite experience to its core tourist elements, coding a stretch of land that encompassed Yosemite Valley and Yosemite Falls. Guests could hike the valley trail and gaze up at the gushing water and majestic but unreachable mountain range. More than just a visual spectacle, the virtual Yosemite included a distinctive soundscape of relaxing music mixed with "real" animal recordings, facilitating a full sensory experience. The park included its own ranger visitor station, where visitors wrote in a guestbook. Virtual hikers signed in and marked their presence. One wrote, "I love these parks, so relaxing"; another noted, "Just beautiful."[24]

As a list of code, the virtual Yosemite, like the virtual Walden, was ultimately programmed to look a certain way and behave according to a limited range of variables. Rather than a recreation of a living, breathing ecosystem, it was a piece of carefully programmed digital artistry and stylized mimicry. Virtual Yosemite was more computer art than natural artifice. Firehawk presented Yosemite as a mostly static experience devoid of movement. Rather than a vibrant ecological reality, it was reminiscent of conventional art, evoking the stillness of *Valley of the Yosemite* (1864) by Hudson River School painter Albert Bierstadt. The National Park Service had spent decades striving to keep Yosemite a "wilderness frozen in time," untainted by human development. Virtual Yosemite achieved a sense of perfection and permanence long quested for. In contrast to the real Yosemite, awash with cars and commotion in summertime, the virtual park provided a welcome sense of tranquility. It afforded a degree of solitude and reflection missing from its real-world twin.

If virtual park visitors desired more active recreational pursuits, they could always head elsewhere. In stark contrast to the restful sounds of Yosemite, *Second Life* Grand Canyon offered machines, noise, and action. It reflected an alternate view of the national park experience as fundamentally about adventure and sport, and connected with the popular reputation of the real-life Grand Canyon for its overflights and white-water rafting. One *Second Life* journalist applauded the range of activities on offer, noting that the sim version had "managed to capture the essence of the Real Life Grand Canyon" as a series of activities, from dirt bikes to balloon rides. In both the real and virtual versions, this park catered to a variety of leisure impulses.[25]

Individual stretches of *Second Life* thus took inspiration from American soil. At Americana and Yosemite, the secondary, digitally coded realm mimicked the external and analogue primary realm. The compatibility of the two worlds and two dimensions reflected the degree to which America itself projected symbols and imagery. In the 1980s, Jean Baudrillard recognized in his travels across the US mainland a utopian experiment so heavy in simulation and symbol that origin or authenticity no longer mattered. America represented a simulated realm, marked by a series of signs, advertising, and imagery.[26] Baudrillard noted the fantasy world of Disneyland and the filmic metropolis of Los Angeles as two perfect examples of places without proper anchors. Coming two decades after Baudrillard's visit, the virtual realm of *Second Life* offered another example of the power of simulation in the American project. As a realm based on code and imagery, *Second Life* represented a virtual perfection of "Symbol America." There seemed little to differentiate the fakery of *Second Life* from the imagery of the greater America.

A Virtual Utopia

In July 1967, German-American philosopher Herbert Marcuse declared the end of utopia. Mass disillusionment accompanied the harmful experiences of the Vietnam War, with both literature and film turning to explore dystopia and discord rather than betterment. The dominant picture of the United States seemed to be one of rising environmental, technological, and social ruin. Ernest Callenbach provided one last vision of utopia in his 1975 novel *Ecotopia*, but, notably, his ecological utopia existed as a breakaway north-western segment of the United States, a new region. Scholar Fredric Jameson, looking back across the tail end of the twentieth century, lamented, "We have seen a marked diminution in the production of new utopias over the last decades."[27] As Nexus Prime, Americana, and other projects took form, *Second Life* came to embody something refreshingly positive and diametrically opposed to such a downward spiral. In their collective efforts, residents set about forging a better way of a life online. Arguably *Second Life* provided a new outlet for aspirational living, a digital utopia.

Prior to *Second Life*, only a few games had explored notions of a digital utopia. In 1981, Don Daglow programmed *Utopia*, a competitive, land-based strategy game for the Intellivision console. Daglow's concept of dueling nations was a precursor to Sid Meier's *Civilization*, a title based on utopian endpoints and linear improvement. In 1991, Gremlin Graphics released *Utopia: The Creation of a Nation* for the Commodore Amiga home computer. The game granted the player godlike powers to shape a new world. Game worlds represented perfect places to imagine new lands and experiment with notions of individual and collective responsibility.

Linden Lab played a crucial role in presenting *Second Life* as a digital utopia. An early Linden advertising campaign encouraged players to "explore, create, compete, connect" in "a new society, a new world, created by you."[28] Linden conceived of its players as citizens in a revolutionary world marked by fundamentally progressive values. With its classical Greek architecture paying homage to the ancient formation of democratic values, Linden Town was a twenty-first-century, virtual utopian project. Finished in 1516, Thomas More's *Utopia* served as both satire and condemnation of England in the sixteenth century as it explored life on a fictional island with different customs. *Second Life* provided a digital version of More's island of Utopia, as well as an experimental space for America 2.0.

Players welcomed the sense of experimentalism. As one resident claimed, "Second Life gives us all a freedom beyond nationality or birth, or conventional perceptions of resources or wealth."[29] The *Washington Post* declared *Second Life* as "in fact, a large-scale experiment in libertarianism."[30] A specific blend of optimism infused the online universe.

Members had a sense of self-importance and missionary zeal. Established in 2007, the *Second Life* Historical Museum collected maps and data pertaining to the online world, and documented the pioneers and ancestors (or primitars) that helped create it.[31] Attention focused on individuals such as Magellen Linden, an early mapper, discoverer, "and somewhat fictional character" who left blogs and artifacts across the digital landscape.[32] *Second Life*'s documenters attributed a sense of manifest destiny to the online enterprise.

Visions of utopia in *Second Life* reflected a techno-cultural movement heavily invested in cyberspace. The machinations of Second Lifers matched the rhetoric of cyberspace idealists such as John Perry Barlow, who welcomed the rise of the World Wide Web, the Internet, and virtual reality technologies. With his "Declaration of Independence," Barlow firmly positioned cyberspace as a new independent territory, distinct and separate from the real world, but still steeped in US terminology and idealism.[33] Barlow hoped that cyberspace could function unconstrained by conventional notions of nation states, laws, corporations, or profit. The digital world represented a chance at a fresh start, unimpeded by corrupt governments and military tyranny. Ambitions for *Second Life* chimed with broader hopes for this new digital revolution. Many early Lifers appeared dedicated to making it a free, creative, and democratic space, and Linden's project became quickly associated with political and intellectual ideas of new virtual communities. Early residents talked revolution and relished the opportunity to carve out a new digital society. Reporter Meadows quoted Columbus's "Following the Light of the Sun, we left the Old World," while others situated *Second Life* as a Shakespearean "Brave New World."[34] *Second Life* showed video games as capable of offering new and revolutionary possibilities, even going as far as forging a new America coded in binary.

The revolutionary energies in *Second Life* flowed from an American tradition. Creators evoked the rhetoric of New World-ism. Talk of a new tolerant society, a refuge for all, and a fresh start seemed not that dissimilar from traditional American ideology at the founding of the country. Like revolutionary soldiers of the late 1700s staking claim to New England, avatars crossed the virtual plains as idealistic citizens. While American settlers superimposed their vision of virgin land onto an indigenous working landscape, in *Second Life*, residents saw a blank map of green pixels and started building. In the 1780s, French-American

J. Hector St. John de Crèvecoeur conceived of an agrarian utopia taking root in the new United States. Watching American society emerge in Orange County, New York, Crèvecoeur pondered, "What then is the American, this new man?" Free from the shackles of European aristocracy, he declared the farmer the "new American."[35] In *Second Life*, residents "tilled the land" in an updated, virtual dimension.[36] Most early Second Lifers focused on transforming their pixel soil into something long-lasting and industrious: the new digital earth yielding new possibilities. For those who felt that America 1.0 was a failed experiment, *Second Life* provided lots of opportunities. Not just an uncharted territory, America 2.0 was an "un-made" place waiting to be created and coded. Without the challenges of disease, climate, or indigenous settlers, the world presented huge possibilities. Only imagination and a Linden fee held back those wanting to create a pure and uncompromised utopia—a new digital society.

Meadows called the Second Life realm "a product of the American Dream" and a "magic kingdom" full of possibilities.[37] The contours of the emergent game world often mirrored twentieth-century utopic visions. Nexus Prime resembled the technological white cities entertained in past World's Fairs and Expositions. Building projects resembled the corporate-utopian town of Fordlandia. *Second Life*'s Commune Utopia paid homage to Aldous Huxley's fictional island of Pala and 1960s counterculture communes such as the Twin Oaks eco-village. One Commune member described his world as "a happy community of hippies, creating the idyllic 60s boho life of music, conversation, love, art, expression of soul."[38]

Without tethers to real-world structures and free from the typical constraints of commercial developments, *Second Life*, more than most fantasy worlds, had the opportunity to "be" exactly what people wanted. Limited only by software, hardware, and collective imagination, projects skyrocketed. Residents felt free to craft their own utopias. The results were personal, idiosyncratic, and often fantastical.

A New American Community

Over the first few years, *Second Life* residents molded and tinkered with their new world. They experimented with social structures, communication forms, and capital accumulation. Communities varied from large residential sectors in Nexus to smaller, remote settlements. Seeking to replicate the intimacy and neighborliness of small-town America, in 2003 Linden itself created the town of Boardman, a zoned sim on the mainland. The planned community preached uniformity of architecture in a world marked by increasing creative chaos. A facsimile of classic 1950s suburbia, Boardman featured identikit households with old-school letterboxes and white picket fences, American flags on porches, and pink flamingoes on well-trimmed lawns. A community hall, functional church, and small shopping center brought residents together. Boardman fully embraced the post-war American dream. It symbolized a Lakewood or Levittown for the twenty-first century and an online refuge from modernity.[39]

In January 2004, Linden employee and community manager Haney Linden, a purple-skinned avatar with a red jumpsuit, publicly listed Boardman (along with a similar project called Brown) as protected virtual spaces. As a "controlled building community," Boardman stayed true to its founding vision. Rules tightly governed local construction, allowing only suburban-style housing, with no possibility for terraforming or modern signage.[40] Selling of wares were restricted to a small stall (or marketplace) in the town center. Statutes limited unwelcome noise and lighting. As a "PG community," Boardman provided a haven for virtual families in a game increasingly sexualized and adult in tone. While land around Boardman continued to change, zoning regulations protected the community from unwelcome intrusion. When in early 2005 the sim fell into disrepair, community manager Jack Linden intervened to actively protect the suburban landscape, arranging the planting of new foliage, the resurfacing of roads, and the construction of a new market. Boardman endured as a shrine to 1950s living—an island of post-war utopia amid a greater virtual project.[41]

The sense of all kinds of communities flourishing in *Second Life* contrasted with reports of life outside and of an America caught in community decline. For some years, social commentators had articulated concern over social decay. In 2000, public policy scholar Robert Putnam declared a crisis in American civic participation and community life. Putnam charted the loss of traditional social structures, from parent–teacher association membership to church attendance, across the latter third of the twentieth century. He concluded that the country had experienced a devastating loss of "social capital" since the 1970s and that an overwhelming sense of "civic malaise" marked society. Where Americans used to bowl in clubs and enjoy each other's company, now they bowled alone.[42]

In his exploration of American community, Putnam was wary of the effect of the Internet and video games on the concept of social capital. According to Putnam, the Internet threatened community health by inviting mutual distrust and misrepresentation, and it fundamentally suffered from a quantifiable lack of "social embeddedness." Putnam felt that "building trust and goodwill is not easy in cyberspace," with many users simply following their own selfish interests and failing to engage with a healthy diversity of thought (a dangerous process called "cyber-balkanization"). Putnam discounted video games as games of chance marked by an unhelpful "solitary nature." New technology, according to Putnam, undermined the social fabric. Atari's 1979 *Bowling* game was just another example of a nation playing alone.[43]

Putnam underestimated the complexity of the relationship between gaming and community. He failed to track the emergence of new subcultures and behaviors evolving in malls and arcades in the 1970s, or the bonding taking place in basements dedicated to *Pong* or hacking. He ignored the growth in highly social home gaming in the 1980s and 1990s, based on localized competitive play, with players competing in bouts of *Mario Kart* and *Street Fighter*. On a broader level, Putnam failed to explore the shift in social capital from the traditional material realm to a new digital domain. Games held the potential to bring people together as much as to keep them apart. They offered novel solutions to the challenge of modern civic engagement. They brought people together through both competitive and collaborative play. Released in 2006, Nintendo's Wii console highlighted the new social turn possible with gaming, as families across the United States (and world) embraced a range of physically based, innovative game titles. With the software release of *Wii Sports* (2006), people came together to play tennis, golf, and ten-pin bowling, but from the comfort of the home.

Arguably the growth of gaming represented a coping mechanism to tackle the very loss of social capital and societal decline charted by Putnam. Games met the problem of growing disillusionment with "life outside" by offering a "life inside." In the 1930s, Americans watched Hollywood B-movies and Mickey Mouse cartoons as a reprieve from the Great Depression. Games offered a similar escape from American reality from the 1970s on. Disneylands of the virtual age, games provided Americans with magic kingdoms to disappear into. By the 2000s, the problems of US society included rising homelessness, a resurgence of racism, and the mass fear of terrorism. Through gaming, players transcended real-life problems; they escaped America 1.0 and made new friends.

As a software title designed to foster community formation and social interaction, *Second Life* was emblematic of this trend. *Second Life* provided community-building projects, democratic action, and dance halls. It offered an array of social capital structures in a virtual domain. Linden's title spoke to significant trends and transitions in contemporary society. It catered to groups disillusioned with real-life systems of support and to those suffering from the ill-effects of long-term civic disengagement. The game attracted those unable or unwilling to physically or mentally join the sorts of organizations Putnam charted the decline of. It catered to those unhappy with their bodies, offering a chance for personal confidence and liberation through a tailored online avatar, akin to a second being. As Phylis Johnson noted, "For some, *Second Life* allows for a much fuller real life than the reality that sometimes imprisons them in daily life."[44] Rather than fakery and facsimile, authenticity marked the new *Second Life* experience. As Meadows explained, "We are more inclined to reveal ourselves when we use our avatars."[45] Players felt authentically themselves. Rather than a temporary escape, the *Second Life* community felt that they were having an authentic and real experience. The game world fulfilled desires for better homes, better cars, and other personal possessions unmet in the real world. It re-empowered the disempowered. *Second Life* responded to the failures of community in First Life. It provided a sense of Americans working through their problems.

Courtesy of its open world design, *Second Life* provided space for diverse types of community. It offered players the chance to form conversations and alliances unfettered by real-world class, gender, or racial dynamics. It encouraged interaction, bonding, and organizing based on the sharing of codes, mods, and programming, alongside conversations at virtual bars over the meaning of life. The creation of a new kind of virtual capital followed animation subroutines and radical community-building projects. It promised a sense of "new" community. *Second Life* represented an experiment in virtual capital. Rather than being evidence of "civic malaise," the title showed just the opposite to be true of Americans: the huge organizing energies of early residents demonstrated a people desperate for connection. In its dynamic first years, *Second Life* indicated a potential rehabilitation project for Putnam's America and an online answer to real-world community decay. It offered a new sense of community, an online America subtly different from the one offline. It offered hope for a revival in community spirit and acted as an engine of social capital. At its best, the virtual project came to embody a revitalization of American freedom (both personal and political) and a revival of democracy and liberal mindedness.

Liberty, Free Play, and Citizenship

Notions of utopia, democracy, and community came together in *Second Life* through two core elements: free play and citizen activism. Play emerged as central to online activity. Linden promoted the online world as a recreational haven. Advertisements centered on the leisure aspects of the digital society. A 2003 Linden press release encouraged players to "start your second life" at $14.95 a month, inviting new residents to join a "remarkable, rich 3D society where you can dance the night away with friends, explore exotic cities, battle in underground tunnels, or sit quietly in your backyard and watch the sun go down over the hills."[46] Linden promoted its new world as one boundless romantic adventure. Presented with a strange amalgam of exotic travel and home comforts, residents soon found themselves exploring new fantastical worlds while entering random dance competitions. The sense of free play also led to transgression. Gambling enterprises proliferated. Early worlds included the Pleasure Cove Sex Resort. Programming mods included Sex and Love 2.0, which allowed players to simulate sexual intercourse. The online world provided a sanctuary for nonconformists. Addictions could be safely pursued in cyberspace, with sexual taboos in particular actively welcomed. The growing popularity of transgressive behavior inside *Second Life* contrasted with the conservatism of American culture outside.

The great liberal experiment encouraged people to play with their identities. In traditional psychology, the concept of the "second life" referred to events past middle age. Self-help literature employed the term as a route toward personal reinvention. The concept of a second life suited the notion of starting afresh in cyberspace. Americans already played with their digital identities through social media interfaces such as Facebook, but *Second Life* allowed a far greater sense of modification. In *Second Life*, people chose completely new identities. The choice of avatars provided the opportunity to be "anyone" and to play any archetype.[47] Players chose outfits, smiles, and body shapes, molding their avatars into digital second selves. Increasingly intimate identification with one's avatar encouraged feelings of absorption, empowerment, vividness, and telepresence. Players simultaneously felt both themselves and their new character. The endless in-game possibilities for tinkering exaggerated the sense of an idealized self. *Second Life* provided a second life.

Second Lifers also embraced the political freedoms and citizen opportunities afforded by the new world. Alongside the playfulness and reckless abandon, virtual citizens assumed responsibility for patrolling their new world. Linden Lab encouraged a participatory fabric from the start. Employees answered individual requests and held town hall meetings over substantive issues. True to the phrase "The new town square is the Internet," *Second Life* emerged as a platform for the engaged netizen.[48] Some residents employed the virtual stage to breathe fresh life into existing real-world activities and struggles. In 2006, Second Life Left Unity (SLLU) formed and soon had its own shop, solidarity area, protest stalls, and campaigns, including ones tied to Gaza.[49] Positioned within a virtual game space, the leftist group challenged real-world capitalism. The existence of SLLU spoke to the rise of unity as well as subversion in the virtual domain. *Second Life* served as an innovative platform for political statement and activism. It offered a stage from which to criticize First Life politics and history. The sense of subversion informed a range of activities, including in-game artistry. In May 2013, Moon Called Lands Native American virtual village sold a "Founding Fathers Native American Painting" in the Linden marketplace for twenty-five Linden dollars.[50] The picture invited new owners to "show your friends where your roots are" by displaying the "greatest chiefs in Native American history" superimposed on Mount Rushmore heads. The painting served as a digital equivalent of the Crazy Horse Monument in Dakota, a real-life protest sculpture started in the 1970s designed to challenge traditional American conceptions of leaders and history.

The degree to which Second Lifers embraced their role of netizens occasionally had repercussions for Linden itself. In 2004, Linden introduced a revised tax system that targeted those with significant virtual property. In response, those affected banded together in protest against the corporation. In defense of the large-scale sim of Americana, Busch's team launched their own anti-Linden campaign. They envisaged their stand as a new American revolution. Paralleling their suffering with that of real-world American colonists at the hands of the British, member Crow explained, "In RL ... when taxes got out of control, our forefathers in Americana dumped a bunch of tea into the harbor, which led to revolution."[51] Crow and his allies called on other residents to "Stop by Americana, see the members, join the revolution," and uttered the rallying cry "Down with the Mad King George." In protest rhetoric, Linden assumed the role of King George and virtual land taxes equated to the sugar tax. Americana devotees saw themselves as patriots defending the founding principles of the new republic. Second Lifers defended the democratic new world against any oppressors. Team member Crow playfully paralleled the cyber-protest with the 1960s civil rights era, in particular Martin Luther King, Jr.'s Birmingham march. Outside commentators took umbrage at the tenuous linkages being made, with one observer noting, "The civil rights movement was significant. A bunch of people in a computer game putting on a show to protest a play-money economy isn't."[52] The Americana revolt failed. *Second Life* version 1.1 introduced new tax costs and marked the demise of large-scale design projects.

The Consumer Republic

Through a variety of tax, membership, and exchange systems, Linden pioneered a new virtual economy. Residents created their own market by designing virtual clothes, character animations, and sim builds, and then selling them to other players. The introduction of Linden dollars in 2004, an in-game currency with US dollar value, linked the fledgling economy to outside financial rewards. Linden oversaw an in-game revolution as thousands of players set about selling their digital creations. As activities coalesced around the design and selling of digital wares, the emergent society increasingly resembled a consumer republic.

In May 2006, Anshe Chung was featured on the cover of *Businessweek* magazine.[53] Chung was a *Second Life* millionaire. That year, Linden's gross domestic product rose to around $64 million. *Second Life* boasted a consumer economy larger than that of a small country. Initially slow to recognize the virtual market, real-world companies rushed into the game world. Disney, Amazon, Microsoft, IBM, NBC, and Nissan all established virtual stores. By 2009, 1,400 real-world companies had a virtual presence.[54] The in-game marketplace featured millions of items, from virtual real estate to fast cars, from animal pets to designer bodies. Individual users crafted all kinds of artisan and bespoke products. Almost everything in the real world re-materialized as a consumer object for sale. *Second Life* existed as one giant department store experience: a total transactional fantasy in which everything that could be seen or interacted with, from foliage to body parts, was for sale.

In some ways, the focus on virtual consumption in the 2000s tracked with the consumer heyday of 1950s America. Steven Miles traces the advent of American consumer culture to affordable Ford cars and working-class spending in the 1910s, but the full consumer revolution only emerged with unbridled access to products in the post-war era.[55] In the 1950s, consumerism emerged as a way of life for the masses. With the latest cars, credit cards, televisions, and property for sale, Americans re-engineered the American Dream into a consumer daydream. The sense of naïve embracement of all products and appliances in the 1950s could be seen again in the mass excitement over a new virtual economy in the 2000s. Rather than the televisions, refrigerators, and automobiles of the post-war consumer boom, *Second Life* consumers bought avatars, skins, and animations. The novelty of shopping proved potent. As Mike Molesworth explains, "much videogame play may also be seen as re-enchanted shopping," with games such as Linden's akin to visits to the mall.[56]

Separated from the world outside by a simple log-in, much of the content of the new Linden marketplace took its inspiration from developments in the twenty-first-century real-life consumer economy. In-game fashion tastes reflected the latest magazine advertisements in *Glamour* and *Vogue*. With everyone a potential model, vanity consumption dominated the market; virtual hair salons were commonplace. *Second Life* sellers tracked contemporary themes of consumption but also emphasized what worked best in the virtual economy. The body, in particular, became a focus of attention. Residents sold avatar skins and facial mods, a virtual equivalent to plastic surgery. With everything changeable and upgradeable, residents made and remade themselves by buying new skins, hair, and outfits. In *Second Life*, the body existed as a consumer product, something to buy and sell in a virtual, post-body world.

Second Life also tapped into a deeper psychological function of consumption. The focus on the body nurtured vanity, narcissism, and identity-construction. Peter Lunt and Sonia Livingstone note that, in the real world, "mass consumption infiltrates everyday life not only as economic processes, social activities and household structures, but also at the level of meaningful psychological experience—affecting the construction of identities, the formation of relationships, the framing of events."[57] In the virtual life, the process seemed even more transparent, with residents talking openly about new faces and new bodies. Such virtual consumption represented a form of personal empowerment. For Miles, "the essence of consumerism . . . lies in the feeling that as consumers we are all gaining some semblance of authority over the everyday construction of our lives *through* consumption."[58] In *Second Life*, consumers exercised authority over every aspect of living, from face paint to home decor. The online world provided a highly immersive and satisfying consumer experience. As Albert Lin noted, "Virtual Worlds offer opportunities, experiences, and pleasures that satisfy many of the basic motivations that drive modern consumption."[59]

Along with a range of tablet and phone games in the 2010s based on buying in-game upgrades (dubbed "pay to win"), *Second Life* charted the course of the new consumer economy. While Internet shopping and the Amazon era created a virtual marketplace, people still received physical goods through the post. Games such as *Second Life* pointed the way toward a fully functional virtual economy bound to the trade of images and symbols—an age of post-material materialism. As Colin Campbell explained, by its increasing reliance on symbols and invisible transactions, "the spirit of modern consumerism is anything but materialistic."[60] For Miles, America had become a nation caught in "desirous daydreaming."[61] *Second Life* highlighted a future in which the physical disappeared and people instead bought fantasies to download.

The Death of Second Life

By the 2010s, the *Second Life* experience seemed in trouble. Statistics indicated a failing economy and a significant reduction in visitor numbers. A range of popular locations disappeared, while much of the hubris surrounding the "new world" dissipated. Reliant on code that was almost a decade old, in graphical terms Linden's title looked antiquated. The search for utopia had also come up short. While Linden tackled problems of gambling and casinos in 2008, online crime and overt sexual content still marred the game world. Rather than a utopic space, *Second Life* seemed closer to a failed experiment.

TRANSGRESSION, THE DARK WEB, AND THE FAILURE OF LIBERAL SPACE

A visit to *Second Life* in the 2010s was a shocking experience. With a significant loss of residents and the demise of familiar, real-world companies, the transgressive aspect of Linden's title became increasingly conspicuous. The most popular entertainment zones were for adults only. Advertisements for sex could be found across many spaces, while nude male avatars greeted the uninitiated in public "welcome" areas. In its lewd, sexualized advertising, *Second Life* resembled the real-life transgressive capital, Las Vegas. Avatars quested for business like insistent ticket touts on the Vegas Strip. Virtual brothels and lap dances proliferated. A virtual Sin City, *Second Life* pushed an "anything goes" philosophy that extended to "furries" sex. People joined precisely for the same reasons that that they journeyed to Las Vegas: to pursue the usually forbidden and the transgressive.

Transgression also filtered into criminality. In 2004, reports surfaced of teenage prostitution rings in *Sims Online*.[62] Fears of taboo sexual conduct consistently marked the *Second Life* landscape. As *Second Life* became more dystopic, the so-called dark web threatened to infiltrate to greater depths. "Age play" games bordered on pedophilia. Games based around "pay/play" rape and sex hunts pushed the boundaries of acceptability. The anonymity of avatars increasingly seemed like a hindrance. As Michael Gerson described *Second Life*, "this experiment suggests that a world that is only a market is not a utopia. It more closely resembles a seedy, derelict carnival—the triumph of amusement and distraction over meaning and purpose."[63] In June 2009, Linden banished sex businesses to their own "adult continent" of Zindra.[64] The act of segregation suggested transgression had, at the least, geographic limits, but somehow only created a common play space, as well as a sense of a sexualized world, or "third life," within the game.

THE UNFULFILLMENT OF VIRTUAL CAPITALISM

The virtual consumer republic also had its problems. The invasion by real-world companies in 2009 faltered as many Second Lifers preferred their own native-bred economy to famous brands. Virtual and real-life capitalism emerged as subtly different goods. Finding little traction, corporate sponsors withdrew. The virtual marketplace continued, offering a distinctive blend of home-grown consumer outfits and avatar accessories. To afford their purchases, residents repeated menial tasks for tiny amounts of Linden currency (at its worst, searching desperately for "free" coins across the game space). Addicted to in-game purchases, virtual reality pioneers became slaves to a new virtual consumer economy. *Second Life* seemed a corporate experiment gone wrong. As one *Second Life* developer, Spider Mandala, stated, "I had ideals that we were going to forge the free idealistic Utopia, but . . . money makes things happen."[65]

Linden's great project demonstrated the convergence between video game content and the worst habits of a consumer economy. A journalist for the *Washington Post* lamented the excesses of *Second Life*, stating, "There are strip malls everywhere, pushing a relentless consumerism."[66] The game amounted to little more than a facile consumer experience, with residents caught in a continual process of searching for and buying new items, a virtual "click and collect." In writing about the first-person shooter *Destiny* (2014), Simon Parkin for the *New Yorker* critiqued how games "often take on the systems of the culture in which they are created, in this case, late capitalism."[67] In the case of *Destiny*, players focused on purchasing and collecting upgrades for their character, mostly armor and weapons, along with an invented, game-specific consumer item: Glimmer. Like many games, *Destiny* made players "work" for "rewards" (the rewards, in turn, making the work easier or at least more interesting). The relationship acted like a closed loop. In *Second Life*, "work" took the form of designing things to sell; residents then used their accrued Linden dollars to buy the work of others. Of course, nothing was really purchased; all of the items remained in-game, image and surfeit. The completeness of the process actually revealed the incompleteness of consumer life. As Theodor Adorno and Max Horkheimer articulated back in the 1940s, mass culture feeds false psychological needs and breeds both citizen passivity and capital growth.[68] The growth of video games appeared a larger victory for distraction, the inauthentic, and the unreal. Games amounted to part of a fake society nurturing its own fake needs.

Games revealed, at least to some, the banal emptiness of the consumer experience. With little else to do but buy and manage home furniture, Will Wright's *The Sims* laid bare the workings of the consumer-capital world. Once the novelty of filling their world with goods wore off, players wondered what to do next. *The Sims* inadvertently highlighted the psychological emptiness of a consumer utopia and the false promise of fulfillment through goods. By its similar focus on designer sims, *Second Life* highlighted the same limits of consumption. For Guy Debord, modern society was marked by spectacles of distraction taking us away from the real.[69] *Second Life* at best was a powerful distraction. A virtual world navigated purely by "looking" and "seeing," *Second Life* offered not much more than image. While impressed by the detail of the virtual recreations, journalist Barry Collins moaned, "Once you've wandered around and seen the sights, what else is there to do?"[70] For Collins, a world like *Second Life* "ultimately serves no purpose." Rather than illustrating the potential rewards of a virtual future, *Second Life* instead highlighted the limits of a world resting on consumer symbols.

COMMUNITY DISSOLUTION

At *Second Life*'s ten-year mark, the incompleteness of the new America proved staggering. A cohesive society, an America 2.0, had failed to materialize. The grand hopes of early residents had given way to a world of vacant malls and cheap sex. Wagner James Au called *Second Life* a "tragedy of commons."[71] Business writer Karyne Levy described a "strange second life" dedicated to sex and dancing.[72] While some diehard players remained, the virtual community consisted mostly of pleasure-seeking individuals. *Second Life* was also no longer American. Around 600,000 players visited the world, but only 40 percent of them originated from the United States. A new internationalism could be witnessed in multiple languages in communication text boxes. However, the greater sense of community had long passed. Pioneers moved on to new software and new projects. Collins simply declared, "It's desolate, its dirty."[73]

Linden's *Second Life* represented a virtual ruin of the early-twenty-first century. The great experiment had unraveled. Nexus Prime had passed from prototype white city to apocalyptic dereliction. Empty boutiques were occupied solely by virtual mannequins. Half-coded, unfinished architectural projects littered the landscape. Collins called *Second Life* a "digital ghost town," the experience simultaneously revealing the failures of utopian communities in both the nineteenth and twenty-first centuries.[74] Wandering the deserted fantasy realm left players with the feeling of being the last survivors in a great utopian experiment wiped out by a mystery contagion. Linden's world was primarily a lonely and unfulfilling experience. *Second Life* had given way to a silent life.[75]

A few wanderers—virtual archaeologists—searched for the ground zero of the great utopian project. They made their pilgrimage to Linden Mansion, all that remained from the original alpha test-scape of *Linden World*, built in 2002. There they found the ruins of a twenty-first-century Jamestown, a frontier ghost town, in the form of a bare shell of a building. Once inside, they made their way downstairs to an old exhibition documenting the Linden project and a time capsule, akin to the Westinghouse capsule at the New York World's Fair, with the promise of rewinding time back to the heyday of virtual civilization, but broken and no longer functional. Just off the virtual coast, the Linden "Man," now half-submerged in virtual water, was reminiscent of the Statue of Liberty sunk and buried in the dystopian movie *Planet of the Apes* (1968). It barely survived as an object from a lost code and a lost golden era. A few colorful offerings could be seen surrounding the Man, paying homage to the oldest relic of the alpha settlement and to the rise and fall of the great virtual experiment.

Conclusion: Converging Worlds

IN 1999 Konami released a survival-horror video game set in the fictional American town of Silent Hill. As Harry Mason, an "everyman" character waking up from a car crash, the player searches for his newly missing daughter, Cheryl, in an unfamiliar urban territory. Mason finds clues scattered across the landscape. The game functions as a cross between a detective novel and a horror movie. With its weird lighting and distorted shapes, the town of Silent Hill is reminiscent of a crooked, creepy ghost house from an old amusement park. Konami's survival-horror game is about the tantalizing nature of mystery and the acute sense of danger felt in navigating a realm marked by malevolence.

As a Japanese software title conceived by Keiichiro Toyama for an American market, *Silent Hill* became a decidedly global product. It offered a vision of small-town America heavily influenced by Konami's other survival-horror series, *Resident Evil* (1996+); the US horror film *Jacob's Ladder* (1990), which tackled the hallucinations of a Vietnam veteran; and the surreal, dream-like world of *Twin Peaks* (1990–1), created by American director David Lynch. The character of Cheryl went by the name of Dolores in earlier versions of the title, taken from Russian-American author Vladimir Nabokov's *Lolita* (1955). Nabokov himself wrote as an outsider to American culture looking in. The in-game "America" of *Silent Hill*, similar to many other Americas crafted by Japanese and European designers, was something peered into, framed from a distance—a voyeuristic experience. It was an uncanny world, not quite real or unreal.[1]

The America that Mason wanders is something dark, ruined, and lost. "This place is like a ghost town," Mason remarks as he first enters Silent Hill. The town is blanketed by fog, providing an additional sense of suffocation and engulfment. The classic American idyll, the small town idealized by countless nineteenth-century writers, is near-abandoned, poisoned, and diluted. It is in decline and on the edge, quite literally: in two of the four possible endings, rocks fall and the otherworldly town collapses. Exploring the imagined past of small-town America, examining its structures and traditions, Mason performs the role of an archaeologist. He searches for meaning, life, and family in the urban and psychological dereliction. He finds animalistic beasts, blood, and horror.

In the film version of Konami's title, scriptwriter Roger Avary envisioned Silent Hill as the actual American town of Centralia, Pennsylvania, a mining settlement rendered uninhabitable by toxic subterranean fires.[2] In May 1962, a landfill fire spread to deserted coal mines beneath Centralia. Only in the 1980s was the scale of the problem realized, when a gas station owner found that his underground tanks were heating gasoline to 172 degrees Fahrenheit and a twelve-year-old boy fell into a steaming sinkhole in his backyard. Over the next decade, the vast majority of residents were relocated, and the town was condemned. The sense of Centralia, a lost town of America, polluted and forlorn, was resurrected in the film version of *Silent Hill* (2006). Caught in a fractured and polluted realm, Mason (depicted by a female lead in the film) searches desperately for Cheryl, as well as a way out from the hell hole. Mason is looking for escape from an evil and broken America.

Silent Hill represents just one of many simulated American experiences in games. From the 1970s onwards, video games promoted a variety of fictions about the American past, present, and future. In their depiction of colonial settlements and futuristic cities, games forwarded different digital states of America to explore. Black-and-white arcade screens, with their pixelated stick figures, imparted binary worlds, free of complexity and nuance, with an "America" that is more symbolic and suggested than tangible and authentic. Video game technology and design progressed, to the point where aesthetic realism became expected in the 2010s, and the ability of games to offer sophisticated narratives became realized. Players began to immerse themselves in fictional worlds arguably more attractive than the analogue world outside.

Past State

Games offered immersion in an imagined past, of an America marked by revolution, military conflict, and the glories of the frontier. Explaining his vision for *Red Dead Redemption*, Rockstar's Dan Houser commented in 2010 that "Westerns are about place . . . and you're talking about a medium, video games, the one thing they do unquestionably better than other mediums is represent geography."[3] Games furnished an alluring geography of the past, bringing to artificial life a range of historic times and places. The *New York Times* called *Red Dead Redemption* a "tour de force" in this regard.[4] Software designers gradually mastered the "look" of American history. Beginning with the basic figures of *Gun Fight*, programmers prioritized aesthetic realism to the point of forging museum-like recreations of the material past, as in Ubisoft's *Assassin's Creed* franchise, a commercial success in the 2010s, that tackled a range of world locations and time periods, including colonial America. *Assassin's Creed* established a convincing veneer of historicity with its replicas and simulacra.

Like other historical fiction, video games set in the American past mostly constructed a story of nation making bound up in notions of sacrifice and heroism. Rather than provide new ways of conceiving historic America, games, in large part, perpetuated familiar narratives. International software companies employed the new technology of computers to preserve an old chimera of American glory days. Akin to "Kings and Queens" European history, titles forwarded a "Generals, Cowboys, and Soldiers" view of American history. Games amounted to a form of interactive nostalgia for tales of white American heroes. In the case of the trans-Mississippi West, computer simulation combined with player agency to re-assert popular, triumphalist meanings of the wild frontier. Aside from educational titles such as *The Oregon Trail* (1971), the commercial game industry fashioned history into a range of missions built around conquest and conflict. First-person shooters such as *Medal of Honor* (1999) morphed drawn-out and horrific world wars into action-packed set pieces. Strategy games such as *Civilization* (1991) perpetuated codes of nation building based on the manipulation of power, ethnicity, and resources, in essence, legitimating the historic "civilizing" of Native Americans, encouraging the exploitation of the environment, and enforcing a clear sense of progress. Games transformed the American past into a new form of interactive entertainment, but one heavily reliant on old comprehensions of history.

Adam Chapman observes that "for most of us imagery and understandings drawn from popular media probably construct the past as much, if not more, than the books of professional historians."[5] Popular media achieves this by a process of technological immersion, with technology (in the form of special effects, sound, and screens) creating a palpable sense of "being there." In the 1970s, video games added a new immersive dynamic of "doing there." Setting them apart from other historical fiction, games involved the player in the fate of the imagined past, encouraging a degree of shared authorship in what Chapman labels the "(hi)story-play-space." Focused on player experience, games forwarded a historical narrative in which the player (at least in appearance) shaped the timeline and determined key events. Gamic depictions of American history focused on the role of the individual, not the masses. American history became a malleable personal fiction, open to different routes, through a digital platform fundamentally about "winning" scenarios and providing the best player experience. Game narratives stretched the historical imagination to include all kinds of outlandish stories revolving around the core character. Titles indulged in troubling meta-fantasies by promoting the idea of a nation founded on an individual player's success at killing, rather than the hard work, democratic values, or migratory forces that shaped real nations. The horror of slavery, the spread of disease, the sacrifice of settlers, the fate of the working-class immigrant—these all disappeared in the pursuit of a more gamic and gratifying past.

Technology and popular history also worked together. In the early twentieth century, action-based Hollywood Westerns helped win the nation over to the format of film by coupling it to a relatively new myth of "cowboys and Indians." That myth soon became dominant in popular understandings of history. In the late twentieth century, video games similarly won over Americans to joystick-based action by coupling digital play to a range of historic victory stories. Computer programs made historic events (such as world wars and the frontier) "gamic" by aligning them with linear missions, kill counts, and decisive wins. Game companies most of all employed the myth of the lone American hero. While film and literature elucidated the fate of fictional leads, video games alone cast the player as the hero. As this project reveals, gamers could take on a number of significant and largely homogeneous archetypes in US-based video games. Players assumed the roles of discoverers and frontier explorers, gods and leaders, cowboys and soldiers, criminals and outlaws, and hardy survivalists. Such gamic roles seemed largely associated with traditional white male characters. They also affirmed some interesting Jungian archetypes, with a notable emphasis on the hero complex/edifice. In the 1990s, British researcher Richard Bartle identified four discernible player types in multiplayer games: achievers, explorers, socializers, and killers.[6] US-set video games mostly abided by the same psychologies. However, above all, the American gamer wanted to become a hero. The effect of this digital heroism has yet to be fully recognized.

Present Lens

Video games also simulated the American everyday. They reflected on the nature of contemporary life. They transformed events, places, and situations into gameplay, action, and entertainment. From small-scale re-creations along specific themes, as in the delivery of beverages in *Tapper*, to sophisticated parodies of managing domestic households, as in the highly popular *The Sims*, titles simulated work, home, and leisure. Games deconstructed, then recoded, American ways of life. In the process, they facilitated the modern corporate condition. Like Happy Meals at McDonald's introducing kids to a world of fast-food convenience and take-away toys, many software companies actively trained young gamers to become consumers. From putting coins in arcade machines in the 1970s to buying upgrades for characters in *Minecraft* (2009) and purchasing microtransactions in the 2010s, games enforced commercial conditioning, reflecting their origins in the mass amusement industry.

Games about the troubled city, the specter of 9/11, and the War on Terror highlighted the extent to which video games reflected and informed serious contemporary debates, but often on the fringes of mainstream society. Players experienced simulations of terror, conspiracy, and even trauma, while the structure of games taught them to respond in coded ways, often based around militarism and aggression. Popular commercial titles articulated a cyber-extension of gun culture, promoting a culture of resolution through violence and a standard way of winning by upgrading weaponry and honing shooting skills. A fictionalized America served as one of several virtual frontlines, or shooting galleries, to face and conquer. With *America's Army* (2002), the US military recognized that many gamers were already soldiers of a kind—and potential recruits for the real frontline. As Elizabeth Losh notes, video games contributed to the growth of the modern military–entertainment complex.[7]

Games equally offered playful, sometimes subversive, commentary on social, political, and cultural issues. Atari's *Missile Command* (1980) critiqued the concept of the winnable Cold War. Rockstar's *Grand Theft Auto* (1997+) series presented a heavily stylized America marked by vice. Such titles celebrated deviance and nonconformity, and cast a critical virtual lens on the present. Situating themselves on the cultural fringes, they parodied US society and its culture industries. Sometimes the technique backfired. In *Super Columbine Massacre RPG!* (2005), Colorado-based independent filmmaker Danny Ledonne explored the Columbine High School killings through the format of a retro-looking, 16-bit video game, tackling themes of media sensationalism, bullying, the motivation of killers, and the influence of gaming on mass culture. Most commentators condemned the title as sensationalist and highly inappropriate, with the ability to "shoot" victims deemed particularly offensive. Understandably perturbed, one parent of a Columbine victim railed, "[It] disgusts me. You trivialize the actions of two murderers and the lives of the innocent."[8] The offensiveness of *Super Columbine* nonetheless paled in comparison to *KKK Mario/Super KKK Brothers 2*, a hacked version of Nintendo's *Super Mario* platform game circulating on the Internet in the same period that had the Italian plumber raising Nazi flags, stomping on African-American babies, and "powering-up" with a white KKK outfit. First-person shooter *Ethnic Cleansing* (2002), another Klan-inspired title, similarly appalled. Gamic subversion sometimes went beyond the parameters of the public good and instead glorified paranoid, xenophobic fantasies of nation.[9]

Noncommercial games emerged in the 2000s as a novel platform for artistic dissidence, environmental concern, and political representation. Media-savvy protest groups employed Flash games as part of wider media campaigns. In December 2010, animal rights group PETA launched *Super Tofu Boy*, a parody of game character *Super Meat Boy* (2010).

Other Flash-based PETA games included Nintendo parodies, the film-inspired *The Pirates of the Carob Bean*, and the Morrissey-inspired *Meat Is Murder: The Game*.[10] Designed to raise awareness of the impact of oil development near the Great Lakes, *Thunderbird Strike* (2017) by indigenous developer and professor Elizabeth LaPensee, tasked the gamer, as a magical thunderbird, with destroying oil pipelines and industrial buildings to save caribou and moose. Republican Senator David Osmek of Minnesota dismissed the game as "an eco-terrorist version of *Angry Birds*."[11]

Republicans and Democrats co-opted games for political campaigning. While computer games had functioned as election simulators for some time (for example, *Campaign 1984* on the Apple II), in the 2016 US election, a number of titles emerged for download on phones and tablets. In *Make America Great Again* (2016), Maverick Game Studio presented Donald Trump as a 1980s Rambo-esque hero who vanquished all foes with bravado and muscle. Anti-Trump titles included *Trumpada* (2015), a platform game in which Trump patrolled the wall between the United States and Mexico, dodging flying tacos while dispatching enemies with his golden combover.[12]

Games in the new millennium increasingly spoke to the problems of contemporary America, with programmers employing dystopian stories and dark satire to highlight the limits of the American mission. Issues of immigration, border control, gun control, and religious radicalism all had their gamic explorations. Commercial titles such as Rockstar's *Grand Theft Auto IV* (2008) highlighted the ailing nature of the American Dream. *GTA IV* related the trials and tribulations of Serbian immigrant Niko Bellic, struggling to make a living in New York City. Another blockbuster title, Ubisoft's *Far Cry 5* (2018), tackled the theme of disillusionment from the perspective of the rural class. The storyline of *Far Cry 5* took place in Montana, where a growing number of disappointed and disenfranchised Americans had turned to The Project at Eden's Gate, a religious doomsday cult led by messiah figure Joseph Seed. With his glasses and slim facial features, Seed closely resembled David Koresh, and the sacrificial fanaticism of Eden's Gate recalled that of Koresh's group, the Branch Davidians, who died during a standoff in Waco, Texas, in 1993. Pitting Seed's followers against locals loyal to the government, *Far Cry 5* highlighted a nation at conflict with itself.

Free from commercial constraints, independent game designers took on issues of race, immigration, and border control with militancy. In 2007, human rights group Breakthrough developed the game *ICED* to highlight the perspective of illegal immigrants facing deportation in New York City. In a story for the *LA Times*, Jamil Moledina, executive director of the Game Developers Conference, noted "how designers have matured and become more politically aware since the days of Pac-Man and Space Invaders."[13] Games amounted to more than just play. The complexities of immigration on a global stage found voice in Lucas Pope's *Papers, Please* (2013), with the player taking on the role of an immigration officer in a fictional Eastern Bloc country, tasked with checking the passports of incoming citizens and hearing their stories. With increasing tensions over US border control allied to the election of Donald Trump, both sides on the US immigration issue employed independent games to make political statements. *Borders* (2017) by Gonzalo Alvarez, a Mexican-American based in Texas, used an overhead action game to show the costs of border incursions, with skulls and skeletons denoting failed crossings, while the *Migrant Trail* (2017) provided touching stories of individual refugees hoping for a fresh start. On the other side of the wall, the Flash game *Border Patrol* (2017) tasked players with shooting Mexicans. As political statements, independent video games pulled few punches.

Along with their proclivity for acerbic wit, video games about contemporary America provided valuable virtual spaces to gain critical perspectives on the analogue world. They highlighted the failures of real society. For designers such as Jane McGonigal, games responded to cravings unanswered in contemporary America.[14] While often pushing an individualist/hero narrative, games also revealed a longing for community. In the 1950s, middle-class Americans belonged to bowling groups and threw Tupperware parties. Fifty years on, the digital age recoded community, refocusing attention on "likes" on Facebook, dancing avatars in online worlds such as *Second Life*, and augmented reality battles. In July 2016, in reaction to a spate of divisive real-life events, including police killings and racially motivated violence, an article in *USA Today* offered video gaming as "exactly the distraction America needs right now."[15] The article referred to the millions of US citizens playing the augmented reality smartphone game *Pokémon Go* (2016), or, as the article put it, "one weird videogame bringing America together." The game furnished an augmented world where cartoon creatures could be seen on top of real-life structures; players could "catch" Pikachu resting on a historic statue, work out their Charizards at coffee shops acting as virtual gyms, and chase illusive Magikarp across beaches. The craze for *Pokémon Go* led to people going outside en masse to hunt Pokémon. Technological magazine *Wired* observed the phenomenon as "weirdly social. Strangers, recognizing what everyone was up to, turned into collaborators or competitors . . . Kids talked to adults. At the end of a week that seemed to focus on people's differences—even fatally so—a game turned at least a few different people into, if not friends, at least people unafraid."[16] Even if contemporary America seemed to be failing and community collapsing, games offered a chance to reconnect in novel ways.

Future America

Game designers equally looked forward in time to offer interactive fictions of America ahead. Program teams created future states of America for the player to explore. In the late 1970s and early 1980s, computer companies presented their products as the first wave of a new digital America, an electronic frontier that infiltrated public and private space alike, and introduced Americans to all kinds of possibilities and potentials. Commodore advertised its microcomputers in the 1980s as in constant use by each family member, showing them how to manage household and business tasks, hobbies and play, and, most of all, how to participate fully in life (rather than just watch television). As Michael Newman contends, games were promoted not just as play, but as "the gateway to the home computer's more transformative uses."[17] Home computers and game machines reflected huge optimism about an unfolding digital nation. Game worlds offered outlines of a new digital society marked by modern heroes, in the guise of David from *WarGames*, the young male computer whizz.

However, even as early as the 1980s, alongside the technological utopias, video games began to map out dystopian futures. Titles such as *Metal Gear, Deus Ex,* and *Half-Life* charted an imagined future very different from that projected at the 1964 New York World's Fair, with its iridescent faith in a new electronic frontier. In contrast to the technological idealism of the World's Fair and early computer advertising, most commercial game titles by the end of the millennium put forward a rhetoric of despondency about the future. They presented a future America in ruins. Rather than a digital future marked by white cities and computer-automated leisure, chaos reigned.

Video games consistently explored the dark side of future America. Titles navigated Cold War and atomic themes in the 1980s, then shifted to depict apocalypses brought on by contagion and environmental collapse in the 2000s. Again, video games explored public worries and concerns. They operated from the cultural fringes to articulate lurking anxieties. Popular titles such as *Resident Evil* (1996) and *Dead Island* (2011) envisaged a nation (and a world) wracked by devastating contagion. Paying homage to Cormac McCarthy's *The Road* (2006), Naughty Dog's *The Last of Us* (2013) portrayed a Texan father and a teenage girl attempting to make sense of a fungus-caused worldwide disaster. In one scene, FEDRA, the gamic equivalent of the Federal Emergency Management Agecy (FEMA), patrolled a flooded, destitute US city, a digital allusion to the decimation of New Orleans caused by Hurricane Katrina, with the player wading in the watery detritus, desperate for escape. In *Horizon Zero Dawn* (2016) by Guerrilla Games, America existed in a state of collapse, with remnants of the past, such as tanks and satellite dishes, lying in a forbidden zone, while robotic creatures resembling animals and dinosaurs patrolled the new wilderness. Games collectively mapped out future doomsday scenarios and offered immersive fantasies about "the end."

The ubiquity of post-apocalyptic game worlds reflected enduring cultural anxiety over impending disasters, doomsdays, and conflict. Game fictions echoed the increasing popularity of dystopian literature and film, and lingering uncertainty over the unquantifiable future. Unlike other media, video games offered the public a chance to actively experience an imagined future and play a unique and decisive role in shaping its contours. Titles such as *Horizon Zero Dawn* and *The Last of Us* placed gamers in the midst of an apocalyptic wilderness and challenged them to prosper. Games offered a powerful cognitive fantasy of heroism and survival. They amounted to a virtual training ground for "video game survivalists" attuned to collecting resources and preparing for a calamitous end of days. They immersed millions of Americans in a battle for future America.

Once again, such battles idealized the lone hero. Coinciding with the action movie *Mad Max: Fury Road* (2015), directed by George Miller, the video game *Mad Max* (2015) by Avalanche Studios threw the player into apocalyptic survival mode. Assuming the identity of Max, a highway patrol officer turned lone survivor, the player navigated a despoiled land, where fuel cans functioned as the new world currency and violent bands marauded and pillaged. Expected to forge a fresh start in the apocalyptic setting, the player set about establishing a small base in a post-American economy revolving around gasoline and gangs. More than just surviving, or eking out a meager existence, the game required the player to fashion a successful livelihood and to dominate over competing factions.

Likewise, in Bethesda's *Fallout 4* (2015), gamers created their own base of operations, including the employ of a robot household servant known as Mr. Handy, in a radioactive America. *Fallout* players fashioned a new America out of old radioactive and polluted dust. They rebuilt the nation piece by piece, atom by atom, code by code. In turn, they re-enacted a national and mythic story of savagery versus civilization. Games explored post-America, but by returning to familiar themes of pre-America. They relied on concepts of frontier justice and vigilantism. Players existed in a neo-Turnerian frontier rendered in digital form. In the majority of post-America titles, the cowboy legend lived on, just in futuristic garb. Games repeated the same process that film underwent in the 1970s, of repackaging the old West of *Outlaw Josey Wales* into space dramas such as *Westworld* and *Star Wars*. Video games cast the player as lead in a new manifest destiny for the twenty-first century.

Alongside the gamer as hero in these digital visions of the future, computers featured prominently in a variety of roles. Video games consistently reflected on the rise of the machine and the symbiotic development of player/human and game/computer. In the strategy title *Fahrenheit 451* (1985), set five years on from the eponymous novel (and with a script co-developed by author Ray Bradbury), books continued to be burned as contraband. The resistance discovered that all book contents had been secretly deposited and preserved as individual microcassettes in a heavily protected New York library. The knowledge of the novel, and of dissidence itself, existed on new media. The player, as fireman Guy Montag, part of a literary underground movement, faced the task of liberating the cassettes for the rebellion and sending them digitally to the outside. Knowledge thus passed from one network to another, with the computer used as a neutral carrier. The good or bad intentions of people, not machines, mattered the most. The future would be determined by Americans.[18]

In other titles, the wayward intentions of computers drove narrative direction. Anxiety over computers fueled the creation of myriad digital game worlds. In titles such as *WarGames* (1984) and *Robotron: 2084* (1982), the ultimate fear was of machines assuming control over the military, society, and culture. In the first-person action adventure *System Shock* (1994), the player as a nameless hacker sought to prevent the powerful artificial intelligence of SHODAN from destroying Earth. Artificial intelligence posed the key threat to human survival. Video game culture also reflected a fear that with the rise of artificial intelligence, people would stop thinking and become automatons trapped in a virtual world. In the movie *Tron*, Dr. Walter Gibbs reassures, "Computers are just machines; they can't think." Alan Bradley replies, "Some programs will be thinking soon." Gibbs retorts, "Won't that be grand? Computers and the programs will start thinking, and the people will stop."

The balance between machine and humanity, game and player, remained of perpetual interest in video game culture between the 1970s and 2010s. From the arcade game of *Berzerk*, with its android stick figures threatening humanity, to the Synths (synthetic intelligence units) resembling humans in *Fallout 3* and *4*, titles explored a future intimately connected to artificial life and simulated reality.[19] Increasingly sophisticated digital realms prototyped the brave new world ahead: America 2.0. Games continually imagined what androids dream of, intertwining the fate of US society with the evolution of computers.

Players thus peered into a new American reality just around the corner: part machine, part artificial intelligence, part player, and part joystick. Games allowed players to enter digital visions of the future and visit science fiction worlds inspired by movies such as *The Matrix* (1999). Virtual reality titles for the Oculus Rift and Sony PSVR delivered all kinds of destinations for players to spend time in. Players could relive past historic traumas, such as 9/11 in *08.46*; wallow in the visual splendor of the Grand Canyon in Immersive Entertainment's *The Grand Canyon VR Experience*; or navigate a future, zombie-infested Wild West in Vertigo's *Arizona Sunshine* (2016). In Samurai Punk's *The American Dream* (2018) for the PSVR, a tongue-in-cheek satire of the nation's gun culture, players could experience the whole of life, from birth to old age, while armed to the teeth.

Games not only peered through the window of a new digital nation; they allowed players to step through the door. Video games facilitated the transition to an alternate American reality. Virtual communities such as *Second Life* encouraged people to adjust to communicating online, having their own avatars, and exploring the possibilities of digital residency. Virtual reality experiences on the PSVR and Oculus Rift introduced players to the physical sensations of virtuality, including, at times, the unfortunate motion sickness. Via extended play on video games, gamers became accustomed to another kind of American reality. Ronald D. Moore's science fiction television show *Caprica* (2010) extrapolated what's next for a growing number of gamers, depicting a world in which teenagers shifted from first playing in a virtual realm (in Caprica, labelled the V-World), to increasingly living in that world, to never being able to return to the material world outside. The history of video games suggests, at the very least, a growing sense of immersion and realism attached to what began as a simple game of *Pong*, with possibly a new digital nation right on America's doorstep.

Digital Nation, Digital Ruins

In the years between 1972, marking the release of *Pong*, and 2016, which witnessed the advent of affordable virtual reality gaming, American society underwent a widespread digital revolution. Embracing the modern computer chip, society witnessed a mass shift toward the digitally imagined landscape. Across four decades, scenes of android war and real-life *WarGames* failed to materialize, but Americans adapted to digital watches, cellphones, tablets, personal computers, and virtual technologies. By the 2010s, society increasingly reflected a fragmented reality, a mixture of material lives and constant occupation of a digital play-space. Hours spent on Facebook, Twitter, and app games replaced hours previously spent watching television and playing outdoor sports such as football. With invisible umbilical cords tying them to Apple iPhones and the Internet, people embraced partly simulated lives. As Timothy Welsh explains, "Ours is a mixed reality, in which big and small screens blend virtual environments into everyday life, and the old binaries dissolve as the virtual and the actual take on a 'strange equality.' "[20]

Video games helped construct this 'digital America' by familiarizing us with new technologies, but more importantly, by tying our vision of America (past, present, *and* future) with the computer chip. Gamic America provided an immersive but transcribed experience of real US architecture, environment, history, and culture, but crucially it advertised and inspired an outline of a new digital nation and powered the digital migration. Games championed the increasing shift to the digital in everyday lives. Online worlds, from *Pong* to *Second Life*, fueled popular desires to inhabit new forms of digital space, while each new technological step brought greater hopes of immersion. Games helped reframe American society around new norms of digital behavior, from typing instructions on computers, to analyzing visual displays, to using ergonomic peripherals.

A Gamer Nation thus helped map out the unfolding digital nation. Games helped us adjust to a simulated American experience. Games such as Atari's *Battlezone* and *Microsoft Flight Simulator* pioneered early forms of simulation. By the 2010s, a simulated version of America included satellite navigation maps, sophisticated military drones, and augmented reality. Alongside more serious technologies, video games helped transform America from its physical, industrial past to something more technological and virtual. Games contributed to a pivotal shift in American life, from "real" to "virtual" consumer life, from physical to electronic social relations, from reading papers to watching Vlogs, and from doorstep politics to online voting. People left behind their department stores, bowling alleys, campaign rallies, and newspapers for Amazon, Pokémon, and fake news.

The rise of a Gamer Nation thus marked a transitional time in the greater American nation, a point at which Americans gradually shifted from visual culture to virtual culture, and from real-world society to an online society. From pixelated stick figures to second lives, gamers embarked on their own digital migration toward this end. Initially designed to offer an escape from the real world, video games instead brought real-world issues into a new digital perspective. Dominant ideologies and patterns of thought transferred across the physical–digital divide. Across the period, computer processing and cultural processing gradually aligned. The CPU arguably became the dual processor of technology and culture by the new millennium. American society entered a state of crossover and blurring, and video games paved the way.

A lot has changed since the 1970s. The distinctive properties of what first marked the Gamer Nation—lengthy software loading times, rudimentary graphics, blips and buzzes, as well as the novelty of playing games on home televisions and at the arcade—seem highly antiquated experiences today. Games now resemble real life, and vice versa. In gamic terms, the "magic circle" may soon disappear, and the uniqueness of game space be lost forever. As Mark Meadows commented on the real–virtual divide in *Second Life*, "the psychological line between them [avatar and self] becomes hazy and thin."[21] Observable differences between avatar/self, online/offline, digital/analogue, unreal/real, and games/life are becoming gradually redundant. Tensions over realism, fantasy, and the uncanny ebb as we move collectively beyond simple binary code and binary ways of thinking. As boundaries become increasingly hard to patrol or even recognize, players "see the virtual as a 'kind' of reality," as Vince Miller observes.[22] In American terms, notions of what marks American space or identity have become more liminal and amorphous.

The inability to differentiate the real from the fictitious seems apposite to a post-truth America of the 2010s marked by conflict. News stories, politics, and even government have become "fake" to some, "real" to others. In gamic terms, driving games influence driving lanes, while video games train the military to operate games of war. A post-truth, post-modern America appears to be a world that blends all kinds of data into a kaleidoscope of competing images and experiences. With *Second Life*, Wagner James Au identified some positive aspects to this symbiosis of real and virtual worlds, noting how money and romance prospered in the virtual realm, only to cross over into the real world, and vice versa. Au called such a process of mobility and symbiosis "mirrored flourishing."[23]

The Gamer Nation, however, is built on precarious code. The mainstream game industry continues to imagine "America" based on simple concepts of what sells—and plays—best. Programmers gamify various experiences, but often without the requisite subtlety. Multiple American voices and stories are lost in the making of commercial gaming products. Like an analogue record turned into an inferior MP3 digital sound stream, much of American life is lost in the binary translation and in the creation of a video game world. Often a mishmash of international authorship, stereotypical characters, and popular cultural references, digital America is rarely a nice place to be or a great example of democratic or utopian values. This gamic nation is also unstable and ephemeral. Without real-world geography, organic roots, or mechanical anchors, gamic America is always open to re-envisioning and reprogramming. Like a Gold Rush town passing from boom to bust, it exists as a place vulnerable to disappearance as much as appearance. Already, a range of gamescapes are deserted or lost in the 2010s, either because corporations pulled the plug or disinterested docents wandered to newer frontiers. Old depictions of gamic America die as arcade machine circuitry gives out. The virtual ruins of *Second Life* resemble a version of America lost; the suburban idyll of Boardman is now deserted and moribund, just an American flag flying in a silent, pixelated world. In *Silent Hill*, despite the illusion of something bigger, the fragments of gamic America never quite fit together. There remain fundamental glitches and omissions in the program of a new America. Video games highlight the hazards of embracing a nation founded purely on computer code.

Introduction: A New Realm of Play

1. Sales figures taken from Gaming History database, https://www.arcade
-history.com/?n=gun-fight-upright-model-no.-597&page=detail&id=1040.

2. Sandomir, "America's Game."

3. Rovell, "NFL TV."

4. Duggan, "Gaming and Gamers."

5. Swanson, "Why Amazing."

6. See 2015 Entertainment Software Association Report, http://www.theesa
.com/wp-content/uploads/2015/04/ESA-Essential-Facts-2015.pdf.

7. Egenfeldt-Nielsen et al., *Understanding Video Games*, 2.

8. "Matt Damon." A representative from the Davie-Brown Index compared
the character recognition of Nintendo's Mario to that of actor Tom Hanks.

9. Newman, *Atari Age*, 2. On Silicon Valley, see Lécuyer, *Making Silicon
Valley.*

10. Huizinga, *Homo Ludens.*

11. Perreault, "Why Do We Love." See also McGonigal, *Reality Is Broken.*

12. Author interview with Nolan Bushnell, 29 May 2018. See "A Nutty
Idea," *They Create Worlds*, Sept 2015, https://videogamehistorian.wordpress
.com/2015/09/03/a-nutty-idea/.

13. W. Smith, "Electronic Games."

14. See "General Info For: Deus Ex Machina," *Spectrum Computing*, https://
spectrumcomputing.co.uk/index.php?cat=96&id=0001373. The game was
released in the United States on the Commodore 64.

15. *Death Race* movie poster (1975). See also Geraghty, *American Science
Fiction*, 53.

16. Blumenthal, "Death Race." See also Kocurek, "Agony."

17. Geist, "Battle for America's Youth."

18. See Montfort and Bogost, *Racing the Beam.*

19. Smith, "Electronic Games."

20. McLuhan, *Understanding Media*, 6.

21. Baudrillard, *America.*

22. Egenfeldt-Nielsen et al., *Understanding Video Games*, 12–13.

23. See Frasca, "Simulation versus Narrative"; Juul, "Games Telling

Stories?"; Fuller and Jenkins, "Nintendo." In *Persuasive Games*, Ian Bogost demonstrated how the video game offered far "more than a book."

24. Crawford and Rutter, "Digital Games," 149; Egenfeldt-Nielsen et al., *Understanding Video Games*, 40; Chapman, *Digital Games as History*, 6; Kocurek, *Coin-Operated Americans*, xvii. See also http://www.playthepast.org; Murkes, "Videogames."

25. In many ways a successor to *Virgin Land*, Richard Slotkin's *Gunfighter Nation* (1991) charted the continued prevalence of symbols and myths in US society. Like Smith, Slotkin noted the endurance of certain imagery, especially the violent frontier, in American popular culture.

26. See Jeffords, *Remasculinization of America*.

27. Jameson, *Raymond Chandler*, 68.

28. Huizinga, *Homo Ludens*, 10.

29. Castronova, *Synthetic Worlds*, 147.

30. Salen and Zimmerman, *Rules of Play*, 95–96.

31. Eco, *Travels in Hyperreality*, 44; Welsh, *Mixed Realism*, 19.

Chapter One: Games and New Frontiers

1. Charles Eames and Ray Eames, *IBM at the Fair* (1964), available at "IBM Pavillion NY World's Fair," *Eames Office*, http://www.eamesoffice.com /the-work/ibm-pavilion-ny-worlds-fair/. For more on Think (1964), see "Think," *Eames Office*, http://www.eamesoffice.com/the-work/think-2/.

2. See "Theme and Symbol." For more on the New York World's Fair, see Cotter and Young, *1964–1965 New York World's Fair*; Samuel, *End of the Innocence*; New York World's Fair 1964–1965 Corporation Records, New York Public Library. For interpretation of the Carousel of Progress, see Allen, "Disneyland."

3. "The New Frontier," acceptance speech of Senator John F. Kennedy, Democratic National Convention, 15 July 1960, John F. Kennedy Presidential Library and Museum, https://www.jfklibrary.org/Asset-Viewer/Archives/JFKSEN -0910-015.aspx.

4. For images and archival sources on the Seattle World's Fair, see Digital Collections, University Libraries, University of Washington, http://digitalcollec tions.lib.washington.edu. For example, the collection includes an image of the bubbleator at http://digitalcollections.lib.washington.edu/cdm/ref/collection /imlsmohai/id/3651. See also the Century 21 Digital Collection, Seattle Public Library, http://cdm16118.contentdm.oclc.org/cdm/landingpage/collection /p15015coll3.

5. *Seattle World's Fair Official Guide Book* (1962), 35; Bell, "Crisis!," 110.

6. IBM archives, http://www-03.ibm.com/ibm/history/.

7. Edwin Newman in the NBC special, "A World's Fair Diary" (1964), broadcast 30 July 1964.

8. As Newman (*Atari Age*, 115) relates, "Home computers at this time were caught between adult and child uses, between seriousness and fun."

9. For early computer game history, see Donovan, *Replay*; Wolf, *Before the Crash*.

10. For more on *Tennis for Two*, see the William Higinbotham Collection 447, Game Studies Collection, Stony Brook University, as well as "The First Video Game?," About Brookhaven, Brookhaven National Laboratory, https://www.bnl.gov/about/history/firstvideo.php.

11. Egenfeldt-Nielsen et al., *Understanding Video Games*, 2; Newman, *Atari Age*, 2.

12. Pursell, *From Playgrounds to PlayStation*, 157.

13. Defense Technical Information Center, "Carmonette Vol. 1" (Nov 1974), 69–70, https://archive.org/details/DTIC_ADA007843.

14. Newman, *Atari Age*, 152. See also Lowood, "Videogames in Computer Space."

15. Pursell, *From Playgrounds to PlayStation*, 1.

16. See Montfort and Bogost, *Racing the Beam*, 109; Interview with Pitfall creator David Crane, *Edge* (Nov 2003); "Atari 2600: Pitfall," *Internet Archive*, https://archive.org/details/atari_2600_pitfall_1983_cce_c-813.

17. Solnit, *River of Shadows*, 215.

18. Brottman, "Ritual, Tension, and Relief."

19. On *Dragon's Lair* and the laserdisc games, see "Player 2 Stage 6: Laser Daze," Dot Eaters—Coin-op Video Game History, available through the Internet Archive Wayback Machine, https://web.archive.org/web/20070723032930/http://www.thedoteaters.com/p2_stage6.php.

20. Kocurek, *Coin-Operated Americans*, 83.

21. Kent, *Ultimate History of Video Games*, 119.

22. Baer, *Videogames*. See also Baer's papers, which are housed at the National Museum of American History, Smithsonian Lemelson Center.

23. Newman, *Atari Age*, 46, 45.

24. Author interview with Bushnell, 29 May 2018, e-mail.

25. Author interview with Bushnell.

26. See D. Winter's collection of ephemera relating to Odyssey, http://www.pong-story.com/odyssey.htm.

27. The first *Pong* machine was a hit at the Andy Capp's Tavern in Sunnydale, CA, with complaints of a broken machine revealing instead an overloaded coin tray.

28. Author interview with Bushnell.

29. Newman, *Atari Age*, 21. See also Kasson, *Amusing the Million*; Wills, *Disney Culture*.

30. Pursell, *From Playgrounds to PlayStation*, 146.

31. Collector sites reveal interesting synergies. For examples of early electro-

mechanical machines, see "Vintage Coin Operated Fortune Tellers, Arcade Games, Digger/Cranes, Gun Games and other Penny Arcade games, pre-1977," http://www.pinrepair.com/arcade/. Sega first released *Duck Hunt* in 1969 as an electromechanical amusement.

32. Newman, *Atari Age*, 200.

33. Author interview with Bushnell; Scott Stilphen, "DP Interviews . . . Ed Salvo," *Digital Press*, http://www.digitpress.com/library/interviews/interview _ed_salvo.html.

34. Author interview with Scott Morrison, 16 March 2018, e-mail.

35. Author interview with Bushnell.

36. Nelson, *American Sports*, 1416–19.

37. Good, "John Kemeny."

38. Kunkel and Laney, "Arcade Alley: Armchair Athletes."

39. A. Michaels, "IGN Presents the History of Madden."

40. Kunkel and Laney, "Arcade Alley: Atari."

41. In 2012, former Nintendo employee Howard Philips published online a 1980s brochure for the Nintendo knitting machine and confirmed a prototype existed.

42. Montfort, *Twisty Little Passages*.

43. Barton, *Dungeons and Desktops*, 29.

44. Wills, "Digital Dinosaurs."

45. For images of *New York! New York!* (Gottlieb 1980), see https://www .arcade-museum.com/game_detail.php?game_id=8858.

46. See *Which State Am I?* (1980), 165 Games, https://archive.org/details /165_Games_oregon_trail; *United States Adventure* (undated), https://archive .org/details/a2_United_States_Adventure_19xx_First_Star_Software_cr_Mr. _Krac_Man.

47. Egenfeldt-Nielsen et al., *Understanding Video Games*, 227.

48. Jenkins, "Game Design"; Bogost, *Persuasive Games*, ix.

49. Robinson, "Out Ran."

50. See Consalvo, *Cheating*.

51. IBM San Antonio 1968 brochure, copy available at http://www.worlds fair68.info/#.

52. Egenfeldt-Nielsen et al., *Understanding Video Games*, 2.

53. Author interview with Don Rawitsch, 16 Feb 2018, e-mail.

54. Brown, *Videogames and Education*, 117. In a 1995 oral history interview, Dale Eugene La Frenz (co-founder of MECC) referred to *The Oregon Trail* as "our flagship product" (p. 37), Charles Babbage Institute Oral History, University of Minnesota.

55. Author interview with Rawitsch. See also Lussenhop, "Oregon Trail"; "I Am Don Rawitsch, a Co-inventor of the Original Oregon Trail Computer

Game. AMA!," *Reddit*, 1 Feb 2016, https://www.reddit.com/r/IAmA/comments
/43ooqf/i_am_don_rawitsch_a_coinventor_of_the_original/.

56. Author interview with Rawitsch.

57. Early survival-horror titles include *Nostrama* (1981) for the Commodore
Pet and *Haunted House* (1982) for the Atari VCS.

58. Author interview with Rawitsch.

59. Parkman, *Oregon Trail*.

60. Gene Fowler, Jr., dir. *Oregon Trail* (Twentieth Century Fox, 1959),
accessed through the Autry Resources Center, Burbank. See also Klausmeyer,
Oregon Trail Stories.

61. Edwards, "9 Myths." For more criticism, see Bigelow, "On the Road."

62. Parkman, "Scenes at the Camp," in *Oregon Trail*, 189.

63. Author interview with Rawitsch.

64. Dyson quoted in Lussenhop, "Oregon Trail."

65. See *Fortune Builder* (1984) manual.

66. Chapman, *Digital Games as History*, 69. See also Edwards, "History of
Civilization."

67. Later iterations of the game focused more on the freedom to generate
distinctive worlds (still bound by the geographic expectations of Earth's
topography/biology).

68. American option (Modern Republic/Democracy) described in Micro-
Prose, *Civilization* PC manual (1991), 129. "In Civilization, Abraham Lincoln
and the Americans are most likely to become a democracy. While they look to
expand, they are not overly aggressive," 135.

69. Sloan, *New World*, 102.

70. Friedman, *Electric Dreams*, ch. 6.

71. Sloan, *New World*, 42.

72. Mir and Owens, "Modelling Indigenous Peoples," 98, 100.

73. Voorhees, "I Play, Therefore I Am," 255.

74. Seabrook, "Game Master."

75. Newspaper advertisement in *La Crosse Tribune*, 21 Oct 1916.

76. Voorhees, "I Play," 257.

77. Douglas, "You Have Unleashed."

78. Poblocki, "Becoming-state," 167, 176.

79. Douglas, "You Have Unleashed."

80. Over time, the *Civilization* series widened its parameters of victory and
success.

81. Turner, "Significance of the Frontier" (1893).

82. Douglas employs Lauren Berlant's phrase "national fantasy" in "You
Have Unleashed."

83. Fuller and Jenkins, "Nintendo," 69.

84. Seattle was originally conceived as a "Festival of the West." See Findlay, *Magic Lands*.

85. Author interview with Storer, 20 Feb 2018. Also see Storer's collection of documents, http://www.cs.brandeis.edu/~storer/LunarLander/LunarLander.html.

86. Edwards, "Forty Years."

87. Montford and Bogost, *Racing the Beam*, 125; Wills, "Wagon Train to the Stars"; T. Smith, "Star Trek."

88. For more on Space Westerns and the frontier idea, see Abbott, *Frontiers Past and Future*; Westfahl, *Space and Beyond*; Katerberg, *Future West*.

89. For a list of the complete contents, see http://www.nywf64.com /weshou08.html. See also Jarvis, *Time Capsules*.

90. Fuller and Jenkins, "Nintendo," 58, 62, 71.

91. Fuller and Jenkins, "Nintendo," 59.

Chapter Two: Playing Cowboys and Indians in the Digital Wild West

1. Roger Ebert's review in the *Chicago Sun-Times*, 1 Jan 1982. This chapter draws from my article "Pixel Cowboys and Silicon Goldmines," published in the *Pacific Historical Review*, 77/2 (May 2008), 273–303.

2. See Kocurek, *Coin-Operated Americans*, 37–65.

3. Kalning, "Ottumwa."

4. Egenfeldt-Nielsen, *Understanding Video Games*, 75.

5. Chris Crawford, "The Atari Years," *Erasmatazz*, http://www.erasmatazz .com/library/the-journal-of-computer/jcgd-volume-5/the-atari-years.html.

6. McQuiddy, "City to Atari."

7. Zak Penn, *Atari: Game Over* (2014, Microsoft).

8. As well as the real-life Calamity Jane's own proclivities for "tall stories," dime novelists employed the character of Calamity Jane to optimum effect, beginning with Edward L. Wheeler's *Deadwood Dick on Deck, or, Calamity Jane, the Heroine of Whoop-Up* (1878).

9. Slotkin, *Gunfighter Nation*, 68–69.

10. Altherr, "Let 'er Rip," 85.

11. White, "Frederick Jackson Turner and Buffalo Bill," 27.

12. For a review of *Buffalo Bill's Rodeo Games*, see *ST Format* (Issue 2), Sept 1989, 70.

13. Smith, *Virgin Land*, 90.

14. Film footage of Cody dates back to 1898, when Thomas Edison recorded Cody firing a rifle. See Carter, *Buffalo Bill Cody*, 424–25.

15. On the West as entertainment landscape, see Aquila, *Wanted Dead or Alive*; Johnson, *New Westers*.

16. On Disney, frontiers, and theme parks, see Steiner, "Frontierland as Tomorrowland"; Francaviglia, "Walt Disney's Frontierland."

17. *Gun Fight* poster (1975).

18. Dykstra, *Cattle Towns*.

19. On the lone hero or last gunfighter/last stand myth, see Cawelti, *Six-Gun Mystique Sequel*, 36–45; Slotkin, *Fatal Environment*, 435–76.

20. Slotkin, *Regeneration through Violence*, 5.

21. In game mechanics, *Blood Bros.* resembled *NAM-1975* (SNK, 1990), a title loosely based on the Vietnam War.

22. White, "Animals," 273, 258.

23. Mystique's other release titles were *Bachelor Party* and *Beat 'Em and Eat 'Em*.

24. Associated Press, "Atari Files"; Morgan, "Custer's Revenge"; "Combatting." Also see Miller, "Indian Group"; and Associated Press, "Indian Movement."

25. King, "Custer's Last Battle," 381.

26. *Captain Fred'k Whittaker, The Dashing Dragoon, or the story of General George A Custer from Wet Point to the Big Horn*, Beadle Boys 3/36 (Dec 1884), 30.

27. Forepaugh Programme, "The Progress of Civilization" (19 July 1887).

28. Ince's *Custer's Last Fight* (101 Bison Film) and materials; Elmer's *Custer's Last Stand* movie brochure and press material. All available at the Autry National Center archives, Los Angeles.

29. Sheridan quoted in Brown, *Bury My Heart at Wounded Knee*, 170–72.

30. Johnson, *New Westers*, 200.

31. See, for example, Clark's difficulty in estimating the vast numbers of "Buffalow, Elk, Antelopes and Wolves" near present-day Billings, Montana; Clark, July 24, 1806, in Thwaites, *Original Journals*, vol. 5, 206.

32. Shaffer, "West Plays West," 376.

33. Shaffer, "West Plays West," 385.

34. Released in the United States in January 2000, *Crazy Taxi* was the first blockbuster title for Sega's Dreamcast console, amassing total US sales of 1.11 million units (the fourth biggest selling title for the Dreamcast).

35. See http://www.virtualoutdoorsman.com/cabelasbiggamehunter3jmp.html; http://www.antlercreeklodge5.homestead.com/home.html. Player review by Carl Cox, 25 Aug 1998, http://www.gamespot.com/sports/deerhunt/review1011e.html.

36. Infogrames Press Release, "Desperados," 2 July 2001 (UK version), http://corporate.infogrames.com/IESA/pressreleases_archive.html.

37. Association for American Indian Development Gun Boycott, www.boycottgun.com; James Brightman, "American Indians Call for Boycott of Activision's Gun," *Game Daily*, 2 Jan 2006, http://biz.gamedaily.com/industry/feature/?rp=39&id=11745.

38. Schiesel, "Way Down Deep."

39. In the lead-up to the game's release, Rockstar keenly promoted *Red Dead Redemption* as part of the "True West." See https://www.rockstargames

.com/newswire/article/2511/the-true-west-history-that-helped-inspire-red-dead
-redemption-ba.html, and https://www.rockstargames.com/newswire/article
/4241/new-inventions-sweep-the-nation-part-one-communication-devices-a
.html, as well as Esther Wright's forthcoming work on the title.

40. Johnson, *New Westers*, 16.

41. Aquila, *Wanted Dead or Alive*, 1.

42. On the frontier process, see Slotkin, *Gunfighter Nation*, 11–12. On
mythic language, see Slotkin, *Fatal Environment*, 18–19.

Chapter Three: Cold War Gaming

1. Brand, "Spacewar"; "Spacewar!," PDP-1 Restoration Project, http://www
.computerhistory.org/pdp-1/spacewar/.

2. Graetz, "Origin of Spacewar."

3. Mullen, "Review of Skylark."

4. Graetz, "Origin of Spacewar."

5. Digital Equipment Corporation (DEC), *1957 to the Present*, brochure
(1978), 3.

6. Leslie, *Cold War and American Science*, 1.

7. Thompson, "Inside the Apocalyptic Soviet Doomsday Machine."

8. On Cold War culture, see Shaw, *Hollywood's Cold War*. On Cold War
gaming, see Halter, *From Sun Tzu to Xbox*, 67–116.

9. Atari Inc., *Coin Connection*, 4/8 (Aug 1980), 2.

10. Theurer quoted in Rubens, "Creation of *Missile Command*."

11. Reagan's "Star Wars" speech, 23 Mar 1983, https://www.reaganlibrary
.gov/research/speeches/32383d.

12. Rubens, "Creation of *Missile Command*."

13. "The Story of Missile Command," part 1, uploaded August 11, 2007,
https://www.youtube.com/watch?v=znq9V-ImHeE.

14. The choice of cities coincided with Atari offices, as programmer Rich
Adam explained: "We were so egocentric that we had missiles coming across
the Pacific aimed at us."

15. "Missile Command," *Gaming-History*, https://www.arcade-history.com
/?page=detail&id=1644.

16. Rubens, "Creation of *Missile Command*"; Atari, *Missile Command*
(1981), VCS manual (author's copy).

17. Oakes, *Imaginary War*.

18. See, for example, Drozdiak, "More than a Million Protest."

19. Kirsch, "Watching the Bombs Go Off," 237, 245.

20. Atari, *Missile Command* (1981), Atari 400/800 manual and box art,
http://www.atarimania.com/game-atari-400-800-xl-xe-missile-command_3441
.html.

21. Atari, *Missile Command*, 400/800 manual.

22. See Weart, *Nuclear Fear*, for a full exploration of nuclear anxieties in the Cold War.

23. Rubens, "Creation of *Missile Command*."

24. Atari, *Missile Command* (1980) arcade machine.

25. Rogers, *Level Up!*, 186.

26. "The Story of Missile Command."

27. Atari's *Combat* (1977), a release title for the VCS, introduced many to competitive gameplay. See also Montfort, "Combat."

28. Moore also compared "the national mania" over video games to the spread of a virus such as herpes. Moore, "Videogames."

29. Maslin, "Red Dawn."

30. For screen shots of *B-1 Nuclear Bomber*, see https://www.mobygames .com/game/b-1-nuclear-bomber.

31. Shapiro, *Atomic Bomb Cinema*.

32. Canby, "*WarGames*."

33. Canby, "*WarGames*."

34. Covert, "High-Tech Hijinks."

35. Coleco Industries, *WarGames* (1984) game manual (author's copy), 13.

36. Jenkins, *Convergence Culture*.

37. Stafford, "WarGames to Screen." See also Montfort and Bogost, *Racing the Beam*, 123–24.

38. The *WarGames* game manual merely describes an "enemy attack," 2.

39. *WarGames* game manual, 2, 10, 13; *WarGames* box art, back cover.

40. Kocurek, *Coin-Operated Americans*, 141, 115.

41. Newman, *Atari Age*, 10.

42. *WarGames* game manual, 2.

43. See Stephens, *Three Mile Island*; Smith, "Soviet Officer"; Schlosser, "World War."

44. An advertisement for *War Room* (1983) by Odyssey/Probe depicted military leaders holding Coleco controllers, with the tagline, "Play the game the generals play . . . for real."

45. Carolipio, "Forgotten Ruins."

46. Kordas, *Politics of Childhood*, 80; McLellan, "Nuclear War Game."

47. Henriksen, *Dr. Strangelove's America*; Kramer, *Dr. Strangelove*.

48. On Mandrake's character, see Case, *Calling Dr. Strangelove*, 73; Cooke, "Dialogue of Fear," 27.

49. Canby, "Documentary."

50. Abalone Alliance, *It's About Times*, May–June 1982, front page (author's copy).

51. New World Computing, *Nuclear War* (1989) Amiga manual (author's copy).

52. Character descriptions in *Nuclear War* manual.

53. On 1980s nuclear fear, see Conze et al., *Nuclear Threats*.

54. See Haraway's classic "Cyborg Manifesto."

55. "Nuclear War" review, *Zap!* 63 (1990), 22.

56. *Amiga Power* 1 (May 1991), 108.

57. Mark Patterson, review, *CU Amiga* (May 1990), 50–51.

58. Oakes, *Imaginary War*, 6.

59. Pemberton, "Why *Fallout 4*'s 1950s Satire Falls Flat." See also Schulzke, "Moral Decision Making in *Fallout*"; Ostroff, "Game After."

Chapter Four: 9/11 Code

1. Although based on a Japanese novel, SquareSoft chose a US location for the game and used a Japanese/American development team.

2. During the attacks, 2,996 lives were lost. See Plumer, "Nine Facts."

3. DeLillo, "In the Ruins of the Future."

4. Stubblefield, *9/11 and the Visual Culture of Disaster*.

5. Stubblefield, *9/11 and the Visual Culture of Disaster*, 3.

6. Gabler, "This Time."

7. Derrida quoted in Stubblefield, *9/11 and the Visual Culture of Disaster*, 3.

8. Gabler, "This Time"; Virilio, *Ground Zero*, 46.

9. Aslan, "Foreword" xii.

10. Stubblefield, *9/11 and the Visual Culture of Disaster*, 3.

11. "Altman says Hollywood."

12. Borradori, *Philosophy*, 108.

13. See chapter five on fighting terrorists in games.

14. Mitchell, *Cloning Terror*, 64.

15. Baudrillard, *Spirit of Terrorism*, 28–29.

16. Baudrillard, *Spirit of Terrorism*, 4.

17. Baudrillard, *Spirit of Terrorism*, 38.

18. Mitchell, *Cloning Terror*, 78.

19. Baudrillard, *Spirit of Terrorism*, 27.

20. Virilio, *Ground Zero*, 68.

21. Dixon, *Film and Television after 9/11*, 30.

22. Engle, *Seeing Ghosts*, 3.

23. Fassone, *Every Game Is an Island*, 57.

24. Wainwright, "New York's Twin Towers."

25. *New York City* cover art, Americana, 1986, http://www.atarimania.com /game-atari-400-800-xl-xe-new-york-city-the-big-apple_6186.html.

26. On eye candy, see, for example, Nitsche, *Video Game Spaces*, 6.

27. Gerstmann, "*Crazy Taxi* 2 Review."

28. More details can be found at http://nintendo.wikia.com/wiki/Little _States.

29. Author interview with Simon Taylor, 21 Feb 2018.

30. *New York Blitz* (1984) cover art / instructions, http://www.ntrautanen.fi /computers/commodore/images/vic/new_york_blitz.jpg.

31. Lagerfeldt post, June 2002, http://www.lemon64.com/forum/viewtopic .php?t=3085&sid=62b2b95addc773369c7a33a14d61f0c1.

32. Microsoft advertisement, *PC Magazine*, 1/8 (Dec 1982), 1.

33. See Corbin, *Base*, 160; Musgrove, "Reality-based Rethinking."

34. Kohler, *Power-Up*, 51; Snider, "Q&A." The New York setting was far more recognizable in Nintendo's *Mario Is Missing* (1992), an educational title in which Mario's brother Luigi searches for lost city artifacts.

35. Michaels, "Save New York Review."

36. *Urban Strike* (1994, video game), "Mission 7: New York City."

37. Ouellette and Thompson, *Post-9/11 Video Game*, 12.

38. Wordsworth, "How *Deus Ex* Predicted the Future."

39. "*Deus Ex* Interview" *Inside Mac Games*, 20 Feb 2003, http://www .insidemacgames.com/news/story.php?ArticleID=7099.

40. Bogost, *Persuasive Games*, 283, 286.

41. Nilges, "Aesthetics of Destruction," 23.

42. Steve Jackson, "Illuminati: The Game of Conspiracy," Steve Jackson Games, http://www.sjgames.com/illuminati/designart.html.

43. Microsoft, *Microsoft Flight Simulator*, Version 3.0 (1988) box art.

44. Baudrillard, *Spirit of Terrorism*, 29.

45. See Wills, "Celluloid Chain Reactions."

46. Ward, "*Blast Corps* Review."

47. Ennis, "Critics Impugned on Trade Center."

48. Page, *City's End*, 199, 190, 189.

49. Žižek, *Welcome to the Desert of the Real*, 16.

50. Baudrillard, *Spirit of Terrorism*, 5, 7.

51. Stubblefield, *9/11 and the Visual Culture of Disaster*, 5.

52. "Sony Delays."

53. Tom Phillips, "How 9/11 Changed Grand Theft Auto 3," 18 Nov 2001, http://www.eurogamer.net/articles/2011-11-18-rockstar-how-9-11-changed -grand-theft-auto-3.

54. Kojima, "Metal Gear Solid 2," 12.

55. Mitchell, *Cloning Terror*, 79.

56. Engle, *Seeing Ghosts*, 12.

57. McCarthy, "Zoolander."

58. DeLillo, "In the Ruins of the Future."

59. Stubblefield, *9/11 and the Visual Culture of Disaster*, 7.

60. Balsamini, "Staten Island." The game was distributed by Uzinagaz, Paris, in 2002. See also https://forums.firehouse.com/forum/the-off-duty-forums/the -off-duty-forums-aa/39638-sad-but-true/page2.

61. Bleiker, "Art after 9/11."

62. Sederstrom, "Artist Axes 9/11 'Invaders' Work." See also Douglas Edric Stanley, "Invaders!"; Abstract Machine, http://www.abstractmachine.net/blog/30-years-of-invasions/.

63. Ehrhardt, "The Division."

64. WDave92 and ZimDoom, Gamespot forum discussion, https://www.gamespot.com/forums/system-wars-314159282/new-york-and-its-destruction-in-video-games-28699314/.

65. Baudrillard, *Spirit of Terrorism*, 52;

66. Stubblefield, *9/11 and the Visual Culture of Disaster*, 145.

67. *Architecture Week* quoted in Stubblefield, *9/11 and the Visual Culture of Disaster*, 125.

68. Mitchell, *Cloning Terror*, 64.

69. Kleinfield, "Creeping Horror." In the *New York Times*, the caption read, "A person falls headfirst after jumping from the north tower of the World Trade Center." See also Junod, "Falling Man."

70. Melnick, *9/11 Culture*, 165.

71. Junod, "Falling Man."

72. "Story Behind the Haunting 9/11 Photo."

73. Žižek, *Welcome to the Desert of the Real*, 13.

74. Junod, "Falling Man"; Cheney, "Life and Death."

75. Stubblefield, *9/11 and the Visual Culture of Disaster*, 55–56, 65.

76. Baudrillard, *Spirit of Terrorism*, 43.

77. The idea of "losing a life" may originate in the multiple balls available in pinball machines, precursors to the modern video game.

78. On perma-death, see Bartle, *Designing Virtual Worlds*, 416; Dixon, "Death."

79. See "The Walk PlayStation VR Simulation," *YouTube*, https://www.youtube.com/watch?v=7m_z2njEaSs&feature=youtu.be.

80. Associated Press, "After Complaints."

81. See "Falling, 2002," Sharon Paz, http://www.sharonpaz.com/falling/.

82. Smith, "Makers of the 9.11 Virtual Reality App."

83. DeLillo, "In the Ruins."

84. See Dillon, "'Disgusting' New Virtual Reality Simulator."

85. See comments on Pretty Neat VR, "8:46 - 9/11 Google Cardboard - SBS 3D - Terrorist Attack Simulator - Controversial VR Experience?," *YouTube*, 13 Sep 2015, https://www.youtube.com/watch?v=HuwHFGPHSIk.

86. Altheide, "Fear, Terrorism, and Popular Culture," 13. Mitchell describes the creation of a virtual memorial by Pastor Antony Mavrakos as one idea for memorializing 9/11, with "Twin Tablets" cast as the Twin Towers and Mavrakos asking people simply to "remember." Mitchell, *Cloning Terror*, 79.

87. Debord, *Society*, thesis one.

88. Sontag, *On Photography*; Stubblefield, *9/11 and the Visual Culture of Disaster*, 28.

89. First published in "Lives Forever Altered," *Bergen County Record*, 12 Sep 2001, 32.

90. Earle, "Photographing the Flag."

91. See https://www.firehero.org/fallen-firefighters/memorial-park/911 -memorial-lift-nation/.

92. On the documentary *9/11*, see Kaplan, "Powerful 9/11 Film Returns."

93. Instructions at Atari Mania, http://www.atarimania.com/game-atari -2600-vcs-towering-inferno_7528.html.

94. Kunkel and Katz, "Arcade Alley: Beyond Science Fiction."

95. *Real Heroes: Firefighter* (Nintendo Wii), Level 4: Yutami High Rise.

96. See http://uk.ign.com/faqs/2010/firefighter-3d-walkthrough-1095780.

97. Miller, "Resistance."

98. Walker, "EA Offers."

99. Bramwell, "Freedom Fighters."

100. Xaosll, Gamespot forum discussion, https://www.gamespot.com /forums/system-wars-314159282/new-york-and-its-destruction-in-video-games -28699314/.

101. Stubblefield, *9/11 and the Visual Culture of Disaster*, 96.

102. Ouellette and Thompson, *Post-9/11*, 37.

Chapter Five: Fighting the Virtual War on Terror

1. Engle, *Seeing Ghosts*, 140.

2. Payne, *Playing War*; Robinson, "Videogames, Persuasion and the War on Terror"; Schulzke, "Virtual War on Terror."

3. DeLillo, "In the Ruins of the Future."

4. See Dixon, *Film and Television after 9/11*, 9; Ouellette and Thompson, *Post-9/11 Video Game*, 9.

5. Spigel, "Entertainment Wars," 129.

6. Voorhees et al., *Guns, Grenades, and Grunts*.

7. Kent, *Ultimate History of Video Games*, 153–55; Jeffords, *Hard Bodies*, 79.

8. Payne, "War Bytes," 272.

9. Annandale, "Avatars of Destruction," 98. See also Hartup, "Why Is It So Appealing."

10. Stern quoted in Gaudiosi, "Modern Warfare 2 Writer."

11. Horiuchi, "Oh My Tech"; Emery, "MPs Row over Modern Warfare Game."

12. Schulzke, "Being a Terrorist," 207.

13. Payne, "War Bytes," 265, 267.

14. Lyman, "A Nation Challenged."

15. Halter, "War Games."

16. Annandale "Avatars of Destruction," 98.

17. "Transcript of President Bush's Address," *CNN*, 21 Sep 2001, http:// edition.cnn.com/2001/US/09/20/gen.bush.transcript/.

18. Dixon, *Film and Television after 9/11*, 1.

19. Annandale, "Avatars of Destruction," 100.

20. Robinson, "Have You Won the War on Terror," 451.

21. Payne, *Playing War*, 2, 28, 31.

22. Hertzberg, "Grinding Axis."

23. Payne, *Playing War*, 10. See also Engelhardt, *End of Victory Culture*.

24. Leonard, "Unsettling the Military Entertainment Complex."

25. Eisiminger quoted in Riddell, "Doom Goes to War."

26. *GamesRadar*, "20 Most Rabidly Patriotic Games."

27. White, "It's a Video Game."

28. Mead, *War Play*, 75. Colonel Casey Wardynski saw *America's Army* as a "virtual soldier experience"; Gonzalez, *Militarizing Culture*, 177. In 2007, the US Army expanded its game/war/public links with the Virtual Army Experience, a travelling simulator.

29. Stahl, "Have You Played the War on Terror?," 112.

30. Virilio, *Ground Zero*, 35.

31. Payne, "War Bytes," 267.

32. Stahl, "Have You Played the War on Terror?," 112.

33. Sample, "Virtual Torture."

34. Herz, *Joystick Nation*, 197.

35. Baudrillard, *Gulf War Did Not Take Place*.

36. Žižek, *Welcome to the Desert of the Real*, 37, 15.

37. Mitchell, *Cloning Terror*, 1.

38. Stahl, "Have You Played the War on Terror?," 112.

39. Delmont, "Introduction," 157.

40. Mitchell, *Cloning Terror*, 98.

41. Mitchell, *Cloning Terror*, 6.

42. Payne, "War Bytes," 271.

43. "September 12th: A Toy World," Games for Change, http://www.games forchange.org/play/september-12th-a-toy-world/.

44. Fassone, *Every Game Is an Island*, 60.

45. Lane, "How 9/11 Affected Games."

46. Said, "Clash of Ignorance"; Baudrillard, "This Is the Fourth War."

47. Annandale, "Avatars of Destruction," 97.

48. Bell-Metereau, "How-to Manual," 156.

49. Žižek, *Welcome to the Desert of the Real*, 9, 12.

50. On Disney's proposed American History theme park, see Perez-Pena, "Disney Drops Plan." On the Enola Gay controversy, see Lilenthal, *History Wars*.

Chapter Six: Grand Theft Los Angeles

1. Cragg, "*Grand Theft Auto* Life-time Sales." For broader work on *GTA*, see Garrelts, *Meaning and Culture of Grand Theft Auto*; Kushner, *Jacked*; Ruch,

"*Grand Theft Auto IV*"; Saklofske, "Thoughtless Play"; Miller, "Accidental Carjack."

2. Acuna, "*Grand Theft Auto* Cost More to Make."

3. Houser quoted in Stuart, "*GTA5* Review." *GTA* explored London as a setting in *Grand Theft Auto: London 1969* (1999).

4. Ulin, *Writing Los Angeles*, xiii.

5. Masters, "This 1897 Film."

6. Banham, *Los Angeles*, 195.

7. Didion, "Pacific Distances," in *After Henry*.

8. Bogost and Klainbaum, "Experiencing Place," 166.

9. Lindsay, *Art of the Moving Picture*, 246, 248.

10. Sorkin, *Exquisite Corpse*, 48–49.

11. Houser quoted in Hill, "*Grand Theft Auto V*."

12. Stuart, "*GTA 5* Review."

13. Garrelts, *Meaning and Culture of Grand Theft Auto*, 12.

14. Sweet, "Idling in Los Santos."

15. Davis, *City of Quartz*.

16. In *Ramona*, Jackson depicted a romantic California society that appealed to American readers.

17. Stuart, "*GTA5* Review."

18. Paget, "Hotel Mistakes."

19. Banham, *Los Angeles*. The game focuses on three characters—Franklin, Michael, and Trevor—who each live in one of the three districts. Franklin lived in South-Central, which resembled Banham's Plains; Michael lived in Beverly Hills, or Banham's Foothills; and Trevor lived in the city's outer reaches, which were similar to Banham's Surfurbia.

20. See, for example, Ruscha, *Every Building on the Sunset Strip*.

21. Sweet, "Idling in Los Santos"; "Science and Poetry."

22. Petit, "City of Angels and Demons."

23. Miller, *Understanding Digital Culture*, 32.

24. Isherwood, *Diaries*, vol. 1, 20.

25. Kirsch, "L.A. without a Map."

26. Houser quoted in Suellentrop, "Americana at Its Most Felonious."

27. Mailer, "Superman Comes to the Supermarket."

28. The term "Californication" dates back to the 1960s, with *Time* magazine in May 1966 suggesting it as an interest of the literary Beat generation.

29. Music video directed by Jonathan Dayton and Valerie Faris.

30. My use of the term "satirical play" derives from the historic use of satire in theater (the satire play), with video games updating the concept to include audience participation.

31. *GTA V* advertisement, Rockstar Games, 2011, https://www.youtube.com/watch?v=QkkoHAzjnUs.

32. See quoted in Ulin, *Writing Los Angeles*, xvii.

33. McWilliams, *Southern California*, 132–33.

34. Davis, *City of Quartz*, 18.

35. McWilliams, *Southern California,* 133; de Beauvoir, *America*, 122.

36. Soja, "Inside Exopolïs," 121.

37. Stuart, "*GTA5* Review."

38. Jameson, *Postmodernism*, 39.

39. Soja, *Postmodern Geographies*, 244.

40. Gildog6 comment on "Raton Canyon: If There's a More Beautiful Spot, I'd Love to See It," r/GrandTheftAutoV, *Reddit*, https://www.reddit.com /r/GrandTheftAutoV/comments/1nxmr5/raton_canyon_if_theres_a_more _beautiful_spot_id/.

41. Mailer, "Superman."

42. Upton quoted in "Uproar Grows"; Thomson quoted in Leung, "Can a Video Game Lead to Murder?"

43. Hubbell, "Nuances of Satire." David Annandale described *GTA: San Andreas* as "a digital incarnation of Mikhail Bakhtin's concept of the carnivalesque." See Annandale, "Subversive Carnival," 89.

44. For more on California Adventure, see Wills, *Disney Culture*, 91.

45. See "Los Disneys: A Game for People who Hate Disney," Tech Crunch, https://techcrunch.com/2007/08/20/los-disneys-a-game-for-people-who-hate -disney/.

46. Baudrillard, *Simulacra and Simulations*, 25.

47. DeVane and Squire, "Meaning of Race and Violence."

48. Whalen, "Cruising in San Andreas," 157.

49. Stuart, "*GTA5* Review."

50. Eco, *Travels in Hyperreality*, 44.

51. Palahniuk quoted in McCracken, *Chuck Palahniuk, Parodist*, 129.

52. Bertozzi, "Marking the Territory," 3.

53. Houser quoted in Hill, "*Grand Theft Auto V*."

54. Newman, *Atari*, 142–43.

55. Delingpole, "The Greatest Joy."

56. Bertozzi, "Marking the Territory."

57. Stuart, "*GTA5* Review."

58. Houser quoted in Hill, "*Grand Theft Auto V*."

59. Ulin, "There It Is."

60. Ellis, *Less Than Zero*, 189–90. See also Baelo-Allue, *Bret Easton Ellis*.

61. Watkins, "Save a Little Water."

62. Chandler quoted in Shoop, "Corpse and Accomplice," 205.

63. Hubbell, "Nuances of Satire."

64. Fine, *Imagining Los Angeles*, 24.

65. Frasca, "Sim Sin City."

66. Higgin, "Play-Fighting," 78.

67. Fine, *Imagining Los Angeles*, 149.

68. Kalning, "What's So Fun About *Grand Theft Auto?*"

69. Hern, "*Grand Theft Auto 5* under Fire."

70. Smith, "Inglorious Nihilism."

71. Wright quoted in Ziff, "I Love LA."

72. White, *Rules of the Game*, excerpt in Ulin, *Writing Los Angeles*, 29.

73. Parkin, "Hunt for One of Gaming's Most Mythical Creatures."

74. Shoop, "Corpse and Accomplice," 206.

75. Smith, "*Grand Theft Auto*."

76. Salter, "Geographical Imaginations of Video Games," 377.

77. Leonard, "Virtual Gangstas," 52.

78. Nakamura, "Race In/For Cyberspace"; Barrett, "White Thumbs, Black Bodies."

79. hooks, *Outlaw Culture*, 178.

80. Lewis, "Yes, It's Misogynistic and Violent."

81. Mulvey, "Visual Pleasure and Narrative Cinema," 11.

82. Gonzalez, "Male Gaze in *GTA V*."

83. Stuart, "*GTA5* Review."

84. Petit, "City of Angels and Demons."

85. Parfitt, "Gamers Petition."

86. Wolfe, "Hey, White People."

87. Examples include *Ape and Essence* (1948), *Chinatown* (1974), and *Blade Runner* (1982). See Davis, *Ecology of Fear*, for further examples.

88. Barrett, "White Thumbs, Black Bodies."

89. Smith, "*Grand Theft Auto*."

90. Rolfe and Lennon, *Bread and Hyacinths*, 23.

91. Bernstein, "How L.A. Noire." Chapman (*Digital Games*, 61) describes *L.A. Noire* as having a "realist simulation style" marked by visual authenticity.

92. Fine, *Imagining*, 119.

93. Davis, *City of Quartz*, 250.

94. The player can choose between missions that kill Trevor or Michael, or a third option in which all core members survive.

95. Cole was himself inspired by a poem by Lord Byron entitled "Childe Harold's Pilgramage." The artist employed verses of the poem in advertisements for his set of paintings. Byron wrote, " 'Tis but the same rehearsal of the past. First freedom and then Glory—when that fails, Wealth, vice, corruption— barbarism last." Birk applied the same concept and sentiment on a local level, to contemporary Los Angeles, in particular the Los Angeles basin and the Holly-wood Hills.

96. Similar in tone is the film still of *Earthquake* (1974) featured in Davis, *Ecology of Fear*, 325.

97. Adamic quoted in McWilliams, *Southern California*, 160.

98. Ulin, *Writing Los Angeles*, xv.

99. Adamic, *Laughing in the Jungle*, excerpted in Ulin, *Writing Los Angeles*, 54.

100. Mencken, *Mencken Chrestomathy*, 291.

101. Itagaki, "Science Fiction and Mysterious Worlds," 382.

102. See quoted in Miranda, "How to Look"; Davis, *City of Quartz*, 20.

103. Houser quoted in Hill, "*Grand Theft Auto*." As outsiders looking in, the British-based Rockstar viewed America with a similar level of disdain to earlier European critics, such as Alexis de Tocqueville.

104. Ulin, *Writing Los Angeles*, xvii.

105. Fine, *Imagining*, 159.

106. Itagaki, "Science Fiction," 377.

107. Houser quoted in Suellentrop, "Americana at Its Most Felonious."

108. Smith, "Inglorious Nihilism."

109. Stuart, "*GTA5* Review."

110. Kalning, "What's So Fun."

111. Scimeca, "Grand Theft Auto."

112. Clinton quoted in Peterson, "Hillary Clinton's History with Video Games."

113. "Police: Man Arrested."

114. Redmond, "Grand Theft Video," 104; Dormans, "The World Is Yours."

115. Smith, "Inglorious Nihilism."

116. McDonald, "*Grand Theft Auto*'s 10 Defining Moments."

117. Huxley, "Los Angeles."

Chapter Seven: Second Life, *Second America*

1. Companies included Apple, while news media representatives from CBS, NBC, and BBC reported events there. Democratic candidate Mark Warner entered *Second Life* as part of his campaign for the 2008 presidential election. For other interpretations of *Second Life* in Game Studies, see Au, *Making of Second Life*; Boellstorff, *Coming of Age in Second Life*; Meadows, *I, Avatar*.

2. Rosedale, "Glimpse Inside a Metaverse," Google Tech Talks, 1 Mar 2006, http://www.allreadable.com/68357nm1.

3. Morehead, "Cybersociality," 180; Au, *Making of Second Life*, 31.

4. Caillois, *Man, Play, and Games*.

5. See "Linden World," Second Life Wiki, http://secondlife.wikia.com/wiki/Linden_World. See also Andrew Linden, "History of Second Life: Linden World Aug 2001," *YouTube*, posted 14 Feb 2011, https://www.youtube.com/watch?v=BbxWXbHoJ2g.

6. See F. Randall Farmer, "Habitat Anecdotes," Fall 1988, http://www.crockford.com/ec/anecdotes.html; Morningstar and Farmer, "Lessons of Lucasfilm's *Habitat*."

7. Steen et al., "What Went Wrong with *The Sims Online*"; Paulk, "Signifying Play."

8. Debord quoted by Micah White, "Gamifying Activism," https://www.micahmwhite.com/gamifying-activism/. There were a number of guides published to help with the transition to living in a virtual world, including Tapley, *Designing Your Second Life*.

9. Nintendo Virtual Boy advertisement, *Computer Gaming World* (Sep 1995), 8–9.

10. Au, *Making of Second Life*, 31–33.

11. Barlow, "A Declaration of the Independence of Cyberspace," Electronic Frontier Foundation, 1996, https://www.eff.org/cyberspace-independence.

12. Au, *Making of Second Life*, 143, xiii.

13. Meadows, *I, Avatar*, 7–8.

14. Cooper, "CyberFrontier and America."

15. Rheingold, *Virtual*, 2.

16. Maiberg, "Why Is 'Second Life' Still a Thing?"

17. Gartner, "Gartner Says 80 Percent of Active Internet Users Will Have A 'Second Life' in the Virtual World by the End of 2011," news release, 24 Apr 2007, https://www.gartner.com/newsroom/id/503861.

18. Au, *Making of Second Life*, 57.

19. Yi, "Online Game Bets on Self-Expression"; "Proposed Community Americana," *Second Life* Forums Archive, 1 Jun 2003, http://forums-archive.secondlife.com/8/16/2852/1.html.

20. Au, *Making of Second Life*, 137.

21. Runte, *National Parks*.

22. Thoreau, *Walden*, 53.

23. In 2017, Tracey Fullerton developed an open world video game (*Walden, a game*) based on Thoreau's time at Walden Pond to much critical acclaim.

24. Firehawk's Yosemite Valley Visitor Center, http://world.secondlife.com/place/4a4fafc4-2801-75f6-aa3d-e2a82dee659b.

25. Diesel, "Sights and Famous Landmarks"; "Grand Canyon," *Second Life* Destination Guide, http://secondlife.com/destination/grand-canyon.

26. Baudrillard, *Simulacra*, 12.

27. Jameson, "An American Utopia," 1.

28. Linden Lab, *Second Life* Advertisement (Sep 2003), https://www.youtube.com/watch?v=KHH2CAE9Y6o. See also Linden Lab, "*Second Life* Opens Public Beta," press release, 28 Apr 2003, https://www.lindenlab.com/releases/second-life-opens-public-beta.

29. Quoted in Au, *Making of Second Life*, 114.

30. Gerson, "Where the Avatars Roam."

31. See http://secondlife.com/destination/sl-historical-museum.

32. See http://secondlife.wikia.com/wiki/Magellan_Linden.

33. See note 11.

34. Meadows, *I, Avatar*, 7. See Laris, referencing Shakespeare, "O Brave New World."

35. Crèvecoeur, *Letters from an American Farmer*, section III.

36. Hirschorn, "Closing the Digital Frontier." The Whole Earth 'Lectronic Link lauded the "neo-Jeffersonian idea of the digital pioneer as a kind of virtual sodbuster" (quoted in Hirschorn).

37. Meadows, *I, Avatar*, 8.

38. See https://community.secondlife.com/forums/topic/188126-commune -utopia-hippie-community-boho-village/.

39. See http://secondlife.com/destination/boardman-themed-community.

40. See "Rules for Boardman, Brown, and De Haro," *Second Life* forums archive, 21 Jan 2004, http://forums-archive.secondlife.com/16/3e/8903/1.html.

41. See http://secondlife.wikia.com/wiki/Boardman.

42. Putnam, *Bowling Alone*, 25.

43. Putnam, *Bowling Alone*, 176, 104.

44. Johnson, *Second Life*, 234.

45. Meadows, *I, Avatar*, 36.

46. Linden Lab, "Start Your Second Life in June 2003," press release, 14 May 2003, https://www.lindenlab.com/releases/start-your-second-life-in-june-2003.

47. The word "avatar" is taken from a Sanskrit word meaning "godly creation."

48. Scott, "My Activist Second Life."

49. On Second Life Left Unity, see http://world.secondlife.com/group /9ecfb08f-75e1-49f0-db5b-ec4dbda0f7bd.

50. Founding Fathers Native American Indian Painting, *Second Life* Marketplace, https://marketplace.secondlife.com/p/Founding-Fathers-native -american-indian-painting/113849?id=113849&slug=Founding-Fathers-native -american-indian-painting.

51. See "Linden Loyalists," *Second Life* forums archive, 21 Aug 2003, http:// forums-archive.secondlife.com/18/9a/4188/3.html.

52. Au, *Making*, 122.

53. "Virtual World, Real Money," front cover, *Businessweek* (1 May 2006).

54. Qing, "Businesses Get a Second Life."

55. Miles, *Consumerism*.

56. Molesworth, "First Person Shoppers," 73.

57. Lunt and Livingstone, *Mass Consumption and Personal Identity*, 24.

58. Miles, *Consumerism*, 25.

59. Lin, "Virtual Consumption," 47.

60. Campbell, *Romantic Ethic*, 89.

61. Miles, *Consumerism*, 18.

62. Harmon, "A Real-Life Debate."

63. Gerson, "Where the Avatars Roam."

64. See "Zindra Is the Official Name of the Second Life Adult Continent," *Second Life Update*, 2 June 2009, http://www.secondlifeupdate.com/news-and-stuff/zindra-is-the-official-name-of-the-second-life-adult-continent.

65. Avatar Spider Mandala quoted in Au, *Making of Second Life*, 48.

66. Gerson, "Where the Avatars Roam."

67. Parkin, "*Destiny*'s Unintended Critique."

68. Adorno and Horkheimer, *Dialectic of Enlightenment*.

69. Debord, *Society of the Spectacle*.

70. Collins, "Whatever Happened to Second Life?"

71. Au, *Making of Second Life*, 56.

72. Levy, "Second Life Has Devolved."

73. Collins, "Whatever Happened to Second Life?"

74. Collins, "Whatever Happened to Second Life?"

75. Jamison, "Digital Ruins."

Conclusion: Converging Worlds

1. Character Commentary, *Silent Hill 3 Official Strategy Guide* (NTT, 2003), 2. For broader interpretations of *Silent Hill*, see the work of Ewan Kirkland, including "Gothic."

2. Dekok, *Unseen Danger*.

3. Houser quoted in Schiesel, "Way Down Deep."

4. Schiesel, "Way Down Deep."

5. Chapman, *Digital Games as History*, 33, 12.

6. Bartle, "Hearts, Clubs, Diamonds, Spades." A Bartle's Quotient (or Test) provided a question-based ranking of traits.

7. Losh, "A Battle," 161.

8. Brian Rohrbough quoted in "Columbine Video Game Draws Relatives' Ire," *NBC News*, 17 May 2006, http://www.nbcnews.com/id/12838169/ns/technology_and_science-games/#.WmIJzCOcZsM.

9. King and Leonard, *Beyond Hate*, 111; Welsh, *Mixed Realism*, 2–6.

10. PETA's games are listed at https://www.peta.org/features/games/.

11. "Thunderbird Strike: Controversial Video Game Takes Aim at Oil Industry," *CBC Radio*, 4 Nov 2017, http://www.cbc.ca/radio/unreserved/from-video-games-to-ya-novels-how-indigenous-art-is-evolving-1.4384041/thunderbird-strike-controversial-video-game-takes-aim-at-oil-industry-1.4384559; "Sen. David Osmek: MN Taxpayers Should Not Be Funding Angry Birds for Eco-Terrorists," Minnesota Senate Republican Caucus, 26 Oct 2017, https://www.mnsenaterepublicans.com/sen-david-osmek-mn-taxpayers-not-funding-angry-birds-eco-terrorists/.

12. On Trump games, see Schoon, "Tap That App."

13. Moledina quoted in Gorman, "Immigration."

14. McGonigal, *Reality Is Broken*, 4.

15. Editors, "*Pokémon Go* is exactly the distraction America needs right now." *USA Today.* 11 July 2016.

16. Dreyfus, "The Weekend *Pokémon Go* Took Over America."

17. Newman, *Atari Age*, 147.

18. Garcia, "Stepping into the Story."

19. For more on *Berzerk*, see http://thedoteaters.com/?bitstory=berzerk.

20. Welsh, *Mixed Realism*, 171.

21. Meadows, *I, Avatar*, 8.

22. Miller, *Understanding Digital Culture*, 33.

23. Au, *Making of Second Life*, xv.

REFERENCES

Aarseth, Espen. *Cybertext: Perspectives on Ergodic Literature*. Baltimore, MD: Johns Hopkins University Press, 1997.

Abbate, Janet. *Inventing the Internet*. Cambridge, MA: MIT Press, 1999.

Abbott, Carl. *Frontiers Past and Future: Science Fiction and the American West*. Lawrence: University Press of Kansas, 2006.

Acuna, Kirsten. "*Grand Theft Auto* Cost More to Make Than Nearly Every Hollywood Blockbuster Ever Made." *Business Insider*. 9 Sept 2013.

Adorno, Theodor, and Max Horkheimer. *Dialectic of Enlightenment*. New York: Herder and Herder, 1972 [1944].

Alexie, Sherman. *Ten Little Indians*. New York: Grove, 2003.

Allen, David. "Disneyland: Another Kind of Reality." *European Journal of American Culture*. 33/1 (March 2014).

Altheide, David L. "Fear, Terrorism, and Popular Culture." In Jeff Birkenstein, Anna Froula, and Karen Randell, eds., *Reframing 9/11: Film, Popular Culture and the War on Terror*. New York: Continuum, 2010.

Altherr, Thomas. "Let 'er Rip: Popular Culture Images of the American West in Wild West Shows, Rodeos, and Rendezvous." In Richard Aquila, ed., *Wanted Dead or Alive: The American West in Popular Culture*. Urbana: University of Illinois Press, 1996.

"Altman says Hollywood 'Created Atmosphere' for September 11." *Guardian*. 18 Oct 1981.

Annandale, David. "The Subversive Carnival of *Grand Theft Auto: San Andreas*." In Garrelts, ed., *The Meaning and Culture of Grand Theft Auto: Critical Essays*. Jefferson, NC: McFarland, 2006.

Annandale, David. "Avatars of Destruction: Cheerleading and Deconstructing the 'War on Terror' in Video Games." In Jeff Birkenstein, Anna Froula, and Karen Randell, eds., *Reframing 9/11: Film, Popular Culture and the War on Terror*. New York: Continuum, 2010.

Aquila, Richard. ed. *Wanted Dead or Alive: The American West in Popular Culture*. Urbana: University of Illinois Press, 1998.

Aslan, Reza, "Foreword." In Jeff Birkenstein, Anna Froula, and Karen Randell, eds., *Reframing 9/11: Film, Popular Culture and the War on Terror*. New York: Continuum, 2010.

Associated Press. "Atari Files Suit to Halt Video Games with Adult Themes." 16 Oct 1982.

Associated Press. "Indian Movement Upset by Game Cartridges." 15 Oct 1982.

Associated Press. "After Complaints, Rockefeller Center Drapes Sept. 11 Statue." *New York Times*. 19 Sept 2002.

Au, Wagner James. *The Making of Second Life: Notes from the New World*. New York: Harper Collins, 2008.

Baelo-Allue, Sonia. *Bret Easton Ellis's Controversial Literature: Writing Between High and Low Fiction*. London: Continuum, 2011.

Baer, Ralph. *Videogames: In the Beginning*. Springfield Township, NJ: Rolenta, 2005.

Balsamini, Dean. "Staten Island 9/11 Survivor Disgusted by 'Tasteless' Internet Video Game." *Staten Island Real-Time News*. 1 Jan 2008.

Banham, Reyner. *Los Angeles: The Architecture of Four Ecologies*. Berkeley: University of California Press, 2009 [1971].

Barrett, Paul. "White Thumbs, Black Bodies: Race, Violence, and Neoliberal Fantasies in *Grand Theft Auto: San Andreas*." *Review of Education, Pedagogy and Cultural Studies*. 28 (2006).

Bartle, Richard. "Hearts, Clubs, Diamonds, Spades: Players Who Suit MUDs." *Journal of MUD Research*. 1/19 (1996).

Bartle, Richard. *Designing Virtual Worlds*. San Francisco: New Riders, 2003.

Barton, Matt. *Dungeons and Desktops: The History of Computer Role-Playing Games*. Wellesley, MA: Peters, 2008.

Baudrillard, Jean. *America*. London: Verso, 1988.

Baudrillard, Jean. *The Gulf War Did Not Take Place*. Bloomington: Indiana University Press, 1995.

Baudrillard, Jean. *Simulacra and Simulations*. Ann Arbor: University of Michigan Press, 1994 [1981].

Baudrillard, Jean. *The Spirit of Terrorism and Other Essays*. London: Verso, 2003.

Baudrillard, Jean. "This Is the Fourth War." *Der Spiegel*. 1/1 (Jan 2004).

Beauvoir, Simone de. *America: Day by Day*. Berkeley: University of California Press, 1999.

Bell, Joseph. "Crisis! How Can We Store Human Knowledge?" *Popular Mechanics*. Nov 1962.

Bell-Metereau, Rebecca. "The How-To Manual, the Prequel, and the Sequel in Post-9/11 Cinema." In Wheeler Winston Dixon, ed., *Film and Television After 9/11*. Carbondale: Southern Illinois University Press, 2004.

Bernstein, Joseph, and Dan Nosowitz. "How *L.A. Noire* Rebuilt 1940s Los Angeles Using Vintage Extreme Aerial Photography." *Popular Science*. 17 May 2011.

Bertozzi, Elena. "Marking the Territory: *Grand Theft Auto IV* as a Playground for Masculinity." In David Embrick, Talmadge Wright, and Andras Lukacs, eds. *Social Exclusion, Power, and Video Game Play: New Research in Digital Media and Technology*. Lanham, MD: Lexington, 2012.

Bigelow, Bill. "On the Road to Cultural Bias: A Critique of The Oregon Trail CD-Rom." *Language Arts*. 74/2 (Feb 1997).

Birkenstein, Jeff, Anna Froula, and Karen Randell, eds., *Reframing 9/11: Film, Popular Culture and the War on Terror*. New York: Continuum, 2010.

Bleiker, Roland. "Art After 9/11." *Alternatives: Global, Local, Political*. 31/1 (Jan–Mar 2006).

Blumenthal, Ralph. "'Death Race' Game Gains Favor; But Not with the Safety Council." *New York Times*. 28 Dec 1976.

Boellstorff, Tom. *Coming of Age in Second Life: An Anthropologist Explores the Virtually Human*. Princeton, NJ: Princeton University Press, 2008.

Bogost, Ian, and Dan Klainbaum. "Experiencing Place in Los Santos and Vice City." In Nate Garrelts, ed., *The Meaning and Culture of Grand Theft Auto: Critical Essays*. Jefferson, NC: McFarland, 2006.

Bogost, Ian. *Persuasive Games: The Expressive Power of Videogames*. Cambridge, MA: MIT Press, 2007.

"Books: Nosepicking Contests." *Time*. 6 May 1966.

Borradori, Giovanni. *Philosophy in a Time of Terror: Dialogues with Jurgen Habermas and Jacques Derrida*. Chicago: University of Chicago Press, 2003.

Bramwell, Tom. "Freedom Fighters." *Eurogamer*. 29 Sept 2003. http://www .eurogamer.net/articles/r_freedomfighters_ps2.

Brand, Stewart. "Spacewar: Fanatic Life and Symbolic Death Among the Computer Bums." *Rolling Stone*. 7 Dec 1972.

Brottman, Mikita. "Ritual, Tension, and Relief: The Terror or the Tingler." In Barry Grant and Christopher Sharrett, eds., *Planks of Reason: Essays on the Horror Film*. Lanham, MD: Scarecrow Press, 2004.

Brown, Dee. *Bury My Heart at Wounded Knee: An Indian History Of The American West*. New York: Henry Holt and Company, 1970.

Brown, Harry. *Videogames and Education*. New York: Routledge, 2008.

Bryman, Alan. "The Disneyization of Society." *Sociological Review*. 47/1 (Feb 1999).

Caillois, Roger. *Man, Play, and Games*. Chicago: University of Illinois Press, 1961 [1958].

Campbell, Colin. *The Romantic Ethic and the Spirit of Modern Consumerism*. London: Blackwell, 1987.

Canby, Vincent. "Documentary on Views about Atom." *New York Times*. 17 Mar 1982.

Canby, Vincent. "*WarGames*, a Computer Fantasy." *New York Times*. 3 June 1983.

Carolipio, Reggie. "Forgotten Ruins: The Roots of Computer Role-Playing

Games and New World Computing." *Venture Beat*. 17 Jan 2011. https://
venturebeat.com/community/2011/01/17/forgotten-ruins-the-roots-of
-computer-role-playing-games-new-world-computing/.

Carter, Robert. *Buffalo Bill Cody: The Man behind the Legend*. New York:
Wiley, 2002.

Case, George. *Calling Dr. Strangelove: The Anatomy and Influence of the
Kubrick Masterpiece*. Jefferson, NC: McFarland, 2014.

Cassell, Justine, and Henry Jenkins, eds. *From Barbie to Mortal Kombat:
Gender and Computer Games*. Cambridge, MA: MIT Press, 1998.

Castronova, Edward. *Synthetic Worlds: The Business and Culture of Online
Games*. Chicago: University of Chicago Press, 2005.

Cawelti, John. *The Six-Gun Mystique Sequel*. Bowling Green, OH: Bowling
Green State University Press, 1999.

Chapman, Adam. *Digital Games as History: How Videogames Represent the
Past and Offer Access to Historical Practice*. New York: Routledge, 2016.

Cheney, Peter. "The Life and Death of Norberto Hernandez." *Globe and Mail*.
22 Sept 2001.

Collins, Barry. "Whatever Happened to Second Life?" *Alphr*. 4 Jan 2010. http://
www.alphr.com/features/354457/whatever-happened-to-second-life/page/0/4.

Collins, Karen. *Game Sound: An Introduction to the History, Theory, and Practice
of Video Game Music and Sound Design*. Cambridge, MA: MIT Press, 2008.

"Combatting Tasteless Video." *Daily Oklahoman*. 8 Feb 1983.

Consalvo, Mia. *Cheating: Gaining Advantage in Videogames*. Cambridge, MA:
MIT Press, 2007.

Conze, Eckart, Martin Klimke, and Jeremy Varon. *Nuclear Threats, Nuclear
Fear and the Cold War of the 1980s*. Cambridge: Cambridge University
Press, 2017.

Cooke, Elizabeth. "The Dialogue of Fear in Fear and Desire and *Dr. Strangelove*." In Jerold Abrams, ed., *The Philosophy of Stanley Kubrick*. Lexington:
University Press of Kentucky, 2007.

Cooper, Jeffrey. "The CyberFrontier and America at the Turn of the 21st
Century: Reopening Frederick Jackson Turner's Frontier." *First Monday*.
5/7 (July 2000).

Corbin, Jane. *The Base: In Search of Al-Qaeda, the Terror Network That Shook
the World*. New York: Simon & Schuster, 2002.

Cotter, Bill and Young, Bill. *The 1964–1965 New York World's Fair: Creation
and Legacy*. Charleston, SC: Arcadia, 2008.

Covert, Colin. "High-Tech Hijinks: Seven Curious Teenagers Wreak Havoc Via
Computer." *Detroit Free Press*. 28 Aug 1983.

Cragg, Oliver. "*Grand Theft Auto* Life-Time Sales Hit 250 Million, *GTA 5* and
GTA Online Ships 70 Million Units." *International Business Times*. 3 Nov
2016.

Crawford, Garry, and Jason Rutter. "Digital Games and Cultural Studies." In Jason Rutter and Jo Bryce, eds., *Understanding Digital Games*. London: Sage, 2006.

Debord, Guy. *The Society of the Spectacle*. Kalamazoo, MI: Black and Red, 1970 [1967].

De Crèvecoeur, J. Hector St. John. *Letters from an American Farmer*. 1782.

Davis, Mike. *City of Quartz: Excavating the Future in Los Angeles*. New York: Verso, 1990.

Davis, Mike. *Ecology of Fear: Los Angeles and the Imagination of Disaster*. New York: Holt, 1998.

DeKok, David. *Unseen Danger: A Tragedy of People, Government, and the Centralia Mine Fire*. Philadelphia: University of Pennsylvania Press, 1986.

Delingpole, James. "The Greatest Joy of Playing *Grand Theft Auto V*? It Lets You Give the Finger to the PC Brigade." *Spectator*. 27 Sept 2014.

DeLillo, Don. "In the Ruins of the Future." *Guardian*. 22 Dec 2001.

Delmont, Matt. "Introduction: Visual Culture and the War on Terror." *American Quarterly*. 65/1 (2013).

DeVane, Ben, and Kurt Squire. "The Meaning of Race and Violence in *Grand Theft Auto: San Andreas*." *Games and Culture*. 3/3–4 (July 2008).

Didion, Joan. *After Henry*. New York: Simon & Schuster, 1992.

Diesel, Piers. "Sights and Famous Landmarks in Second Life America." *Second Life Enquirer*. 29 Dec 2013. http://www.slenquirer.com/2013/12/sights-and -famous-landmarks-in-second_29.html.

Dillon, Nancy. "'Disgusting' New Virtual Reality Simulator Allows Players to Experience Horror of 9/11 from Inside North Tower." *New York Daily News*. 2 Nov 2015. http://www.nydailynews.com/news/national/disgusting -virtual-reality-game-lets-players-live-9-11-article-1.2421205.

Dixon, Dan. "Death: A Minor Annoyance or an Invitation to Play?" 2008, unpublished.

Dixon, Wheeler Winston. ed. *Film and Television After 9/11*. Carbondale: Southern Illinois University Press, 2004.

Donovan, Tristan. *Replay: The History of Video Games*. Lewes: Yellow Ant, 2010.

Dormans, Joris. "The World Is Yours: Intertextual Irony and Second Level Reading Strategies in *Grand Theft Auto*." *Games Research* (Oct 2006).

Douglas, Christopher. "'You Have Unleashed a Horde of Barbarians': Fighting Indians, Playing Games, Forming Disciplines." *Postmodern Culture*. 13/1 (2002).

Dreyfus, Emily. "The Weekend *Pokémon GO* Took Over America." *Wired*. 11 Aug 2016.

Drozdiak, William. "More than a Million Protest Missiles in Western Europe." *Washington Post*. 23 Oct 1983.

Duggan, Maeve. "Gaming and Gamers." Pew Research Center. 15 Dec 2015. http://www.pewinternet.org/2015/12/15/gaming-and-gamers/.

Dykstra, Robert. *The Cattle Towns*. Lincoln: University of Nebraska Press, 1983 [1968].

Earle, Harriet. "Photographing the Flag." *Alluvium*. 4/5 (2015).

Eco, Umberto. *Travels in Hyperreality*. London: Picador, 1986.

"If America Is United on One Thing Right Now, It's Pokémon." *USA Today*. 11 July 2016.

Edwards, Benj. "40 Years of Lunar Lander." *Technologizer*. 19 July 2009. http://www.technologizer.com/2009/07/19/lunar-lander/2/.

Edwards, Benj. "The History of Civilization." *Gamasutra*. 18 July 2007. https://www.gamasutra.com/view/feature/1523/the_history_of_civilization.php?page=2.

Edwards, Phil. "9 Myths You Learned from Playing *The Oregon Trail*." *Vox*. 2 June 2015. https://www.vox.com/2015/2/6/7987697/oregon-trail-game-real-life.

Egenfeldt-Nielsen, Simon, Jonas Heide Smith, and Susana Pajares Tosca, eds. *Understanding Video Games: The Essential Introduction*. 3rd ed. New York: Routledge, 2016.

Ehrhardt, Michelle. "*The Division* Doesn't Want You to Think About 9/11." *Kill Screen*. 8 Feb 2016. https://killscreen.com/articles/the-division-doesnt-want-you-to-think-about-911/.

Ellis, Bret Easton. *Less Than Zero*. New York: Simon & Schuster, 1985.

Emery, David. "MPs Row over Modern Warfare Game." BBC News. 9 Nov 2009. http://news.bbc.co.uk/1/hi/8342589.stm.

Engelhardt, Tom. *The End of Victory Culture*. New York: Basic Books, 1995.

Engle, Karen. *Seeing Ghosts: 9/11 and the Visual Imagination*. Montreal: McGill-Queen's University Press, 2009.

Ennis, Thomas. "Critics Impugned on Trade Center." *New York Times*. 15 Feb 1964.

Fassone, Riccardo. *Every Game Is an Island: Endings and Extremities in Video Games*. New York: Bloomsbury, 2017.

Fine, David. *Imagining Los Angeles: A City in Fiction*. Reno: University of Nevada Press, 2000.

Francaviglia, R. "Walt Disney's Frontierland." *Western Historical Quarterly*. 30/2 (Summer 1999).

Frasca, Gonzalo. "Simulation versus Narrative: Introduction to Ludology." In Mark Wolf and Bernard Perron, eds., *Video/Game/Theory*. New York: Routledge, 2003.

Frasca, Gonzalo. "Sim Sin City: Some Thoughts About *Grand Theft Auto 3*." *Game Studies*. 3/2 (Dec 2003).

Friedan, Betty. *The Feminine Mystique*. New York City: W. W. Norton, 1963.

Friedman, Ted. *Electric Dreams: Computers in American Culture*. New York: New York University Press, 2005.

Fuller, Mary, and Henry Jenkins. "Nintendo and New World Travel Writing: A Dialogue." In Steven Jones, ed., *Cybersociety: Computer-Mediated Communication and Community*. Thousand Oaks, CA: Sage, 1995.

Gabler, Neal. "This Time, the Scene Was Real." *New York Times*. 16 Sept 2001.

GamesRadar (US Division), "20 Most Rabidly Patriotic Games." *GamesRadar*. 4 July 2008. http://www.gamesradar.com/20-most-rabidly-patriotic-games/.

Garcia, Cristina. "Stepping into the Story: Players Participate in 'Interactive Fiction.' " *Time*. 125/19. 13 May 1985.

Garrelts, Nate. ed. *The Meaning and Culture of Grand Theft Auto: Critical Essays*. Jefferson, NC: McFarland, 2006.

Gaudiosi, John. "Modern Warfare 2 Writer: 'The Airport Level Was a Risk We Had to Take.' " *Gamepro*. 19 Nov 2009. https://web.archive.org/web /20110207091150/http://www.gamepro.com/article/features/213011 /modern-warfare-2-writer-the-airport-level-was-a-risk-we-had-to-take/.

Geist, William. "The Battle for America's Youth." *New York Times*. 5 Jan 1982.

Geraghty, Lincoln. *American Science Fiction Film and Television*. New York: Berg, 2009.

Gerson, Michael. "Where the Avatars Roam." *Washington Post*. 6 July 2007.

Gerstmann, Jeff. "*Crazy Taxi 2* Review." *Gamespot*. 29 May 2001. https:// www.gamespot.com/reviews/crazy-taxi-2-review/1900-2766530/.

Gonzalez, Diego. "The Male Gaze in *GTA V*." Spring 2016. https://sites.williams .edu/engl117s16/uncategorized/the-male-gaze-in-gta-v/.

Gonzalez, Roberto. *Militarizing Culture: Essays on the Warfare State*. Walnut Creek, CA: Left Coast, 2010.

Gildog6 comment. "Ranton Canyon: If There's a More Beautiful Spot, I'd Love to See It." *Grand Theft Auto V* comments, *Reddit*. https://www.reddit.com /r/GrandTheftAutoV/comments/1nxmr5/raton_canyon_if_theres_a_more _beautiful_spot_id/.

Good, Keith. "John Kemeny and Tecmo's BASIC FTBALL Granddaddy." *TecmoBowlers.com*. 17 Oct 2017. https://tecmobowlers.com/2017/10/17 /kemenys-basic-ftball-tecmos-granddaddy/.

Gorman, Anna. "Immigration Debate Finds Itself in Play." *Los Angeles Times*. 9 July 2007.

Graetz, J. M. "The Origin of Spacewar." *Creative Computing*. 7/8 (Aug 1981).

Halter, Ed. "War Games." *Village Voice*. 12 Nov 2002.

Halter, Ed. *From Sun Tzu to Xbox*. New York: Thunder Mouth Press, 2006.

Haraway, Donna. "A Cyborg Manifesto." In *Simians, Cyborgs and Women*. New York: Routledge, 1991.

Harmon, Amy. "A Real-Life Debate on Free Expression in a Cyberspace City." *New York Times*. 15 Jan 2004.

Hartup, Phil. "Why Is It So Appealing to Play as a Terrorist in Video Games?" *New Republic*. 7 June 2014.

Henriksen, Margot. *Dr. Strangelove's America: Society and Culture in the Atomic Age*. Berkeley: University of California Press, 1997.

Herman, Leonard. *Phoenix: The Rise and Fall of Home Video Games*. Self-published, 1984.

Hern, Alex. "*Grand Theft Auto 5* Under Fire for Graphic Torture Scene." *Guardian*. 18 Sept 2013.

Hertzberg, Hendrik. "Grinding Axis." *New Yorker*. 11 Feb 2002.

Herz, J. C. *Joystick Nation*. London: Abacus, 1997.

Higgin, Tanner. "Play-Fighting: Understanding Violence in *Grand Theft Auto III*." in Nate Garrelts, ed., *The Meaning and Culture of Grand Theft Auto: Critical Essays*. Jefferson, NC: McFarland, 2006.

Hill, Matt. "*Grand Theft Auto V*: Meet Dan Houser, Architect of a Gaming Phenomenon." *Guardian*. 7 Sept 2013.

Hirschorn, Michael. "Closing the Digital Frontier." *Atlantic*. July–Aug 2010.

hooks, bell. *Outlaw Culture: Resisting Representations*. New York: Routledge, 1994.

Horiuchi, Vince. "Oh My Tech. 'Call of Duty' Has Troubling Scene." *Salt Lake Tribune*. 16 Nov 2009.

Hubbell, Gaines. "Nuances of Satire: Falling into *GTA V*'s Biopolitical Trap." *Higher Level Gamer*. 8 Oct 2013. https://higherlevelgamer.org/2013/10/08 /nuances-of-satire-falling-into-gta-vs-biopolitical-trap/.

Huizinga, Johan. *Homo Ludens*. London: Routledge & Kegan Paul, 1944.

Huntemann, Nina, and Matthew Payne, eds. *Joystick Soldiers: The Politics of Play in Military Video Games*. New York: Routledge, 2010.

Huxley, Aldous. "Los Angeles: A Rhapsody" (1926). In *Stories, Essays, & Poems*. London: Dent, 1937.

Isherwood, Christopher. *Diaries*, vol. 1. Katherine Bucknell, ed. New York: Random House, 1996.

Itagaki, Lynn Mie. "Science Fiction and Mysterious Worlds." In Blake All-mendinger, ed., *A History of California Literature*. New York: Cambridge University Press, 2015.

Jackson, Helen Hunt. *Ramona*. Little, Brown, 1884.

Jameson, Fredric. *Postmodernism: Or, the Cultural Logic of Late Capitalism*. Durham, NC: Duke University Press, 1991.

Jameson, Fredric. *Raymond Chandler: The Detections of Totality*. New York: Verso, 2016.

Jameson, Fredric. "An American Utopia." In Fredric Jameson and Slavoj Žižek, eds., *An American Utopia: Dual Power and the Universal Army*. New York: Verso, 2016.

Jamison, Leslie. "The Digital Ruins of a Forgotten Future." *Atlantic*. Dec 2017.

Jarvis, William. *Time Capsules: A Cultural History*. Jefferson, NC: McFarland, 2002.

Jeffords, Susan. *The Remasculinization of America: Gender and the Vietnam War*. Bloomington: Indiana University Press, 1989.

Jeffords, Susan. *Hard Bodies: Hollywood Masculinity in the Reagan Era*. New Brunswick, NJ: Rutgers University Press, 2004.

Jenkins, Henry. *Convergence Culture: Where Old and New Media Collide*. New York: New York University Press, 2006.

Jenkins, Henry. "Game Design as Narrative Architecture." In Noah Wardrip-Fruin and Pat Harrigan, eds., *First Person: New Media as Story, Performance, and Game*. Cambridge, MA: MIT Press, 2003.

Johnson, Michael. *New Westers: The West in Contemporary American Culture*. Lawrence: University Press of Kansas, 1996.

Johnson, Phylis. *Second Life, Media and the Other Society*. New York: Peter Lang, 2010.

Junod, Tom. "The Falling Man." *Esquire*. Sept 2003.

Juul, Jesper. "Games Telling Stories? A Brief Note on Games and Narratives." *Game Studies*. 1/1 (July 2001).

Kalning, Kristin. "Ottumwa, Video Game Capital of the World?" *NBC News*. 6 May 2009. http://www.nbcnews.com/id/30588831/#.WmNg4SOcZsM.

Kalning, Kristin. "What's So Fun About *Grand Theft Auto*?" *NBC News*. 28 Apr 2008. http://www.nbcnews.com/id/24356537/ns/technology_and_science -games/t/whats-so-fun-about-grand-theft-auto/#.Wk9jcCOcZsM.

Kapell, Matthew Wilhelm, and Andrew Elliott, eds. *Playing with the Past: Digital Games and the Simulation of History*. New York: Bloomsbury, 2013.

Kaplan, Don. "Powerful 9/11 Film Returns." *New York Daily News*. 10 Sept 2016.

Kasson, John. *Amusing the Million: Coney Island at the Turn of the Century*. New York: Hill & Wang, 1978.

Katerberg, William. *Future West: Utopia and Apocalypse in Frontier Science Fiction*. Lawrence: University Press of Kansas, 2008.

Kent, Steven. *The Ultimate History of Video Games*. Roseville, CA: Prima, 2001.

King, Captain Charles. "Custer's Last Battle." *Harpers* 81/483 (Aug 1890).

King, Richard, and David Leonard. *Beyond Hate: White Power and Popular Culture*. Burlington: Ashgate, 2014.

Kirkland, Ewan. "Gothic Videogames, Survival Horror, and the Silent Hill Series." *Gothic Studies*. 14/2 (2012).

Kirsch, Adam. "L.A. without a Map." *Slate*. 30 June 2003. http://www.slate.com /articles/arts/culturebox/2003/06/la_without_a_map.html.

Kirsch, Scott. "Watching the Bombs Go Off: Photography, Nuclear Landscapes and Spectator Democracy." *Antipode*. 29/3 (1997).

Klausmeyer, David. *Oregon Trail Stories: True Accounts of Life in a Covered Wagon*. Guilford, CT: Globe Pequot, 2003.

Kleinfield, N. R. "A Creeping Horror." *New York Times*. 12 Sept 2001.

Kohler, Chris. *Power-Up: How Japanese Video Gamers Gave the World an Extra Life*. New York: Dover, 2016.

Kojima, Hideo. "Metal Gear Solid 2: Grand Game Plan." Translated by Mark Laidlaw. Konami Script. 8 Jan 1999.

Kocurek, Carly. "The Agony and the Exidy: A History of Video Game Violence and the Legacy of Death Race." *Game Studies*. 12/1 (Sept 2012).

Kocurek, Carly. *Coin-Operated Americans: Rebooting Boyhood at the Video Game Arcade*. Minneapolis: University of Minnesota Press, 2015.

Kordas, Ann Marie. *The Politics of Childhood in Cold War America*. New York: Routledge, 2013.

Kramer, Peter. *Dr Strangelove or: How I Learned to Stop Worrying and Love the Bomb*. BFI Film Classics. Houndsmill: Palgrave, 2014.

Kunkel, Bill, and Frank Laney. "Arcade Alley: Atari Video Computer System" *Video*. 2/3 (Summer 1979).

Kunkel, Bill, and Frank Laney. "Arcade Alley: Armchair Athletes—Sports, Mattel-Style." *Video*. 12 (Aug 1980).

Kunkel, Bill, and Arnie Katz. "Arcade Alley: Beyond Science Fiction, A New Breed of Games." *Video*. 6/8 (Nov 1982).

Kushner, David. *Jacked: The Outlaw Story of Grand Theft Auto*. Hoboken, NJ: Wiley, 2012.

Lane, Rick. "How 9/11 Affected Games." *Bit-Gamer*. 12 Sept 2011. https://www.bit-tech.net/reviews/gaming/pc/how-9-11-affected-games/3/.

Laris, Michael. "O Brave New World That Has Such Avatars in It." *Washington Post*. 4 Jan 2009.

Lécuyer, Christophe. *Making Silicon Valley: Innovation and the Growth of High Tech, 1930–1970*. Cambridge, MA: MIT Press, 2006.

Levy, Karyne. "Second Life Has Devolved into a Post-Apocalyptic Virtual World." *Venture Beat*. 2 Aug 2014. https://venturebeat.com/2014/08/02/second-life-has-devolved-into-a-post-apocalyptic-virtual-world-the-weirdest-thing-is-how-many-people-still-use-it/.

Lewis, Helen. "Yes, It's Misogynistic and Violent, but I Still Admire *Grand Theft Auto*." *Guardian*. 21 Sept 2013.

Lin, Albert. "Virtual Consumption: A Second Life for Earth?" *Brigham Young University Law Review*. 2/2 (2008).

Lindsay, Vachel. *The Art of the Moving Picture*. New York: Random House, 1915.

Leonard, David. "Unsettling the Military–Entertainment Complex: Video Games and a Pedagogy of Peace." *SIMILE: Studies in Media & Information Literacy Education*. 4/4 (2004).

Leonard, David. "Virtual Gangstas, Coming to a Suburban House Near You: Demonization, Commodification, and Policing Blackness." In Nate Garrelts, ed., *The Meaning and Culture of Grand Theft Auto: Critical Essays*. Jefferson, NC: McFarland, 2006.

Leslie, Stuart. *The Cold War and American Science: The Military-Industrial-Academic Complex at MIT and Stanford*. New York: Columbia University Press, 1993.

Leung, Rebecca. "Can a Video Game Lead to Murder?" *CBS News*. 17 June 2005.

Lilenthal, Edward, ed. *History Wars: Enola Gay and Other Battles for the American Past*. New York: Henry Holt, 1996.

Losh, Elizabeth. "A Battle for Hearts and Minds." In Nina Huntemann and Matthew Payne, eds., *Joystick Soldiers: The Politics of Play in Military Video Games*. New York: Routledge, 2010.

Lowood, Henry. "Videogames in Computer Space: The Complex History of *Pong*." *IEEE Annals of the History of Computing*. (July–Sept 2009), 1–19.

Lunt, Peter Kenneth, and Sonia Livingstone. *Mass Consumption and Personal Identity: Everyday Economic Experience*. Buckingham: Open University Press, 1992.

Lussenhop, Jessica. "Oregon Trail: How Three Minnesotans Forged Its Path." *City Pages*. 19 Jan 2011.

Lyman, Rick. "A Nation Challenged: Hollywood; White House Sets Meeting with Film Executives to Discuss War on Terrorism." *New York Times*. 8 Nov 2001.

Maiberg, Emanuel. "Why Is 'Second Life' Still a Thing?" *Motherboard*. 29 Apr 2016. https://motherboard.vice.com/en_us/article/z43mwj/why-is-second-life -still-a-thing-gaming-virtual-reality.

Mailer, Norman. "Superman Comes to the Supermarket." *Esquire*. Nov 1960.

Marfels, Christian. *Bally: The World's Game Maker*. 2nd ed. Las Vegas: Bally Technologies, 2007.

Maslin, Janet. "Red Dawn." *New York Times*. 10 Aug 1984.

Masters, Nathan. "This 1897 Film Was the First Movie Made in Los Angeles." *KCET*, 2 Oct 2016. https://www.kcet.org/shows/lost-la/this-1897-film-was -the-first-movie-made-in-los-angeles.

"Matt Damon Who? Pac-Man, Mario Are the Heroes." *Reuters*. 16 May 2008. https://www.reuters.com/article/us-videogames-characters/matt-damon-who -pac-man-mario-are-the-heroes-survey-idUSSP16728620080516.

McCarthy, Todd. "Zoolander." *Variety*. 27 Sept 2001.

McCracken, David. *Chuck Palahniuk, Parodist: Postmodern Irony in Six Transgressive Novels*. Jefferson, NC: McFarland, 2016.

McDonald, Kelly. "*Grand Theft Auto*'s 10 Defining Moments." *Guardian*. 13 Sept 2013.

McGonigal, Jane. *Reality Is Broken*. New York: Penguin, 2011.

McLellan, Dennis. "Nuclear War Game Bombs Its Way to Success." *Los Angeles Times*. 18 Apr 1986.

McLuhan, Marshall. *Understanding Media*. London: Routledge, 1964.

McQuiddy, Marian. "City to Atari: 'E.T.' Trash Go Home." *Alamogordo Daily News*. 27 Sept 1983.

McWilliams, Carey. *Southern California: An Island on the Land*. Salt Lake City: Gibbs-Smith, 1946.

Mead, Corey. *War Play: Videogames and the Future of Armed Conflict*. Boston: Houghton Mifflin, 2013.

Meadows, Mark. *I, Avatar: The Culture and Consequences of Having a Second Life*. Berkeley: New Riders, 2008.

Medved, Michael. "Zap! Have Fun and Help Defeat Terrorists, Too." *USA Today*. 30 Oct 2001.

Melnick, Jeffrey. *9/11 Culture*. Chichester: Wiley, 2009.

Mencken, H. L. *A Mencken Chrestomathy*. New York: Vintage, 1982.

Michaels, Al. "IGN Presents the History of Madden." *IGN*. 8 Aug 2008. http://uk.ign.com/articles/2008/08/08/ign-presents-the-history-of-madden.

Michaels, R. J. "Save New York Review." *Ahoy!* 2 (1984). https://archive.org/stream/Ahoy_Issue_02_1984-02_Ion_International_US#page/n51/mode/2up.

Miles, Steven. *Consumerism as a Way of Life*. New York: Sage, 1998.

Miller, Greg. "Resistance: Burning Skies Review." *IGN*. 28 May 2012. http://uk.ign.com/articles/2012/05/28/resistance-burning-skies-review.

Miller, Kiri. "The Accidental Carjack: Ethnography, Gameworld Tourism, and *Grand Theft Auto*." *Game Studies*. 8/1 (Sept 2008).

Miller, Marjorie. "Indian Group to Picket Local Seller of *Custer's Revenge*." *Daily Oklahoman*. 30 Jan 1983.

Miller, Vincent. *Understanding Digital Culture*. London: Sage, 2011.

Mir, Rebecca, and Trevor Owens. "Modeling Indigenous Peoples: Unpacking Ideology in Sid Meier's Colonization." In Matthew Kapell and Andrew Elliott, eds., *Playing with the Past: Digital Games and the Simulation of History*. New York: Bloomsbury, 2013.

Miranda, Carolina. "How to Look at Los Angeles." *Los Angeles Times*. 24 July 2015.

Mitchell, W. J. T. *Cloning Terror: The War of Images, 9/11 to the Present*. Chicago: University of Chicago Press, 2011.

Molesworth, Mike. "First-Person Shoppers: Consumer Ways of Seeing in Videogames." In Mike Molesworth and Janice Denegri-Knott, eds., *Digital Virtual Consumption*. New York: Routledge, 2012.

Montfort, Nick. "Combat in Context." *Game Studies*. 6/1 (Dec 2006).

Montfort, Nick. *Twisty Little Passages*. Cambridge, MA: MIT Press, 2003.

Montfort, Nick, and Ian Bogost. *Racing the Beam: The Atari Video Computer System*. Cambridge, MA: MIT Press, 2009.

Moore, Mike. "Videogames: Sons of *Pong*." *Film Comment*. 19/1 (Jan/Feb 1983).

Moran, Lee. "College Student Steals Truck." *New York Daily News*. 25 Sept 2013.

Morehead, John. "Cybersociality: Connecting Fun to the Play of God." In Craig Detweiler, ed., *Halos and Avatars: Playing Video Games with God*. Louisville, KY: Westminster John Knox Press, 2010.

Morgan, Andy. "*Custer's Revenge* Games Returned." *Daily Oklahoman*. 1 Feb 1983.

Morningstar, Chip, and F. Randall Farmer. "The Lessons of Lucasfilm's *Habitat*." In Michael Benedikt, ed., *Cyberspace: First Steps*. Cambridge, MA: MIT Press, 1990.

Mullen, R. D. "Review of Skylark." *Science Fiction Studies*. 2/3 (Nov 1975). https://www.depauw.edu/sfs/birs/bir7.htm#Smith.

Mulvey, Laura. "Visual Pleasure and Narrative Cinema." *Screen*. 16/3 (1975).

Murkes, Mark. "Videogames and American Society." MA thesis, University of Groningen, Netherlands, 2004.

Murray, Janet. *Hamlet on the Holodeck: The Future of Narrative in Cyberspace*. Cambridge, MA: MIT Press, 1997.

Musgrove, Mike. "Reality-based Rethinking." *Washington Post*. 21 Sept 2001. https://www.washingtonpost.com/archive/business/2001/09/21/reality -based-rethinking/b8ea5e48-d7da-4eb6-834f-6f5b94c35f29/?utm_term =.8047c59dc863.

Nakamura, Lisa. "Race In/For Cyberspace: Identity Tourism and Racial Passing on the Internet." In David Trend, ed., *Reading Digital Culture*. New York: Wiley, 2001.

Nelson, Murry, ed. *American Sports: A History of Icons, Idols, and Ideas*. Santa Barbara, CA: Greenwood, 2013.

Newman, Michael. *Atari Age: The Emergence of Video Games in America*. Cambridge, MA: MIT Press, 2017.

Nilges, Mathias. "The Aesthetics of Destruction: Contemporary US Cinema and TV Culture." In Jeff Birkenstein, Anna Froula, and Karen Randell, eds., *Reframing 9/11: Film, Popular Culture and the War on Terror*. New York: Continuum, 2010.

Nitsche, Michael. *Video Game Spaces: Image, Play and Structure in 3D Worlds*. Cambridge, MA: MIT Press, 2008.

Oakes, Guy. *The Imaginary War: Civil Defense and American Cold War Culture*. New York: Oxford University Press, 1995.

Ostroff, Joshua. "The Game After: A Brief History of *Fallout 4*'s Post-Apocalyptic Retrofuture." *Exclaim!* 15 Dec 2015. http://exclaim.ca/gaming/article/game _after-brief_history_of_fallout_4s_post-apocalyptic_retrofuture.

Ouellette, Marc, and Jason Thompson. *The Post-9/11 Video Game: A Critical Examination*. Jefferson, NC: McFarland, 2016.

Page, Max. *The City's End: Two Centuries of Fantasies, Fears, and Premonitions of New York's Destruction*. New Haven, CT: Yale University Press, 2008.

Paget, Mat. "Hotel Mistakes *GTA 5* Screenshot for Real Los Angeles." *Gamespot*. 29 June 2016. https://www.gamespot.com/articles/hotel-mistakes-gta-5 -screenshot-for-real-los-angel/1100-6441385/.

Parfitt, Ben. "Gamers Petition for Sacking of Gamespot Writer Who Criticized *GTAV* for Misogyny." *MCV*. 18 Sept 2013. http://www.mcvuk.com/articles /media-pr/gamers-petition-for-sacking-of-gamespot-writer-who-criticised -gtav-for-miso.

Parkin, Simon. "The Hunt for One of Gaming's Most Mythical Creatures." *New Yorker*. 26 Aug 2013.

Parkin, Simon. "*Destiny*'s Unintended Critique of Consumerism." *New Yorker*. 2 Oct 2014.

Parkman, Francis. *The Oregon Trail: Sketches of Prairie and Rocky Mountain Life*. 1849.

Paulk, Charles. "Signifying Play: *The Sims* and the Sociology of Interior Design." *Game Studies*. 6/1 (Dec 2006).

Payne, Matthew Thomas. *Playing War: Military Video Games After 9/11*. New York: New York University Press, 2016.

Payne, Matthew Thomas. "War Bytes: The Critique of Militainment in *Spec Ops: The Line*." *Critical Studies in Media Communication*. 31/4 (2014).

Pemberton, Tom. "Why *Fallout 4*'s 1950s Satire Falls Flat." *Atlantic*. 8 Dec 2015.

Perez-Pena, Richard. "Disney Drops Plan for History Theme Park in Virginia." *New York Times*. 29 Sept 1994.

Perreault, Greg. "Why Do We Love Video Games?" *Huffington Post*. 6 Dec 2017. https://www.huffingtonpost.com/greg-perreault/why-do-we-love-video -game_b_4740425.html.

Peterson, Andrea. "Hillary Clinton's History with Video Games and the Rise of Political Geek Cred." *Washington Post*. 21 Apr 2015.

Petit, Carolyn. "City of Angels and Demons." *GameSpot*. 16 Sept 2013. https:// www.gamespot.com/reviews/grand-theft-auto-v-review/1900-6414475/.

Plumer, Brad. "Nine Facts About Terrorism in the United States Since 9/11." *Washington Post*. 11 Sept 2013.

Poblocki, Kacper. "Becoming-State: The Bio-Cultural Imperialism of Sid Meier's *Civilization*." *Focaal: The European Journal of Anthropology*. 39 (2002).

"Police: Man Arrested for Playing Real-Life *Grand Theft Auto*." *Eyewitness News*. 25 Sept 2013. http://abc13.com/archive/9260827/.

Poole, Steven. *Trigger Happy: The Inner Life of Videogames*. London: Fourth Estate, 2000.

Pursell, Caroll. *From Playgrounds to PlayStation: The Interaction of Technology and Play*. Baltimore: Johns Hopkins University Press, 2016.

Putnam, Robert. *Bowling Alone: The Collapse and Revival of American Community*. New York: Simon & Schuster, 2000.

Qing, Liau Yun. "Businesses Get a Second Life." *ZD Net*. 16 Dec 2009. http://www.zdnet.com/article/businesses-get-a-second-life/.

Redmond, Dennis. "*Grand Theft Video*: Running and Gunning for the US Empire." In Nate Garrelts, ed., *The Meaning and Culture of Grand Theft Auto: Critical Essays*. Jefferson, NC: McFarland, 2006.

Rheingold, Howard. *Virtual Community: Homesteading on the Electronic Frontier*. Cambridge, MA: MIT Press, 2000.

Riddell, Rob. "Doom Goes to War." *Wired*. 1 Apr 1997.

Robinson, Martin. "Out Ran: Meeting Yu Suzuki, Sega's Original Outsider." *Eurogamer*. 22 Mar 2015. http://www.eurogamer.net/articles/2015-03-22-out-ran-meeting-yu-suzuki.

Robinson, Nick. "Have You Won the War on Terror? Military Videogames and the State of American Exceptionalism." *Millennium: Journal of International Studies*. 43/2 (2014).

Robinson, Nick. "Videogames, Persuasion and the War on Terror." *Political Studies*. 60/3 (2012).

Rogers, Scott. *Level Up!: The Guide to Great Video Game Design*. 2nd ed. New York: Wiley, 2014.

Rolfe, Lionel, and Nigey Lennon. *Bread and Hyacinths: The Rise and Fall of Utopian Los Angeles*. Los Angeles: Classic Books, 1992.

Rovell, Darren. "NFL TV Viewership Dropped an Average of 8 Percent This Season." *ESPN*. 5 Jan 2017.

Rubens, Alex. "The Creation of *Missile Command* and the Haunting of its Creator, Dave Theurer." *Polygon*. 15 Aug 2013. https://www.polygon.com/features/2013/8/15/4528228/missile-command-dave-theurer.

Ruch, Adam. "*Grand Theft Auto IV*: Liberty City and Modernist Literature." *Games and Culture*. 7/5 (2012).

Runte, Alfred. *National Parks: The American Experience*. Lincoln: University of Nebraska Press, 1982.

Ruscha, Edward. *Every Building on the Sunset Strip*. Los Angeles: Dick de Ruscha, 1966.

Said, Edward. "The Clash of Ignorance." *Nation*. 4 Oct 2001.

Saklofske, Jon. "Thoughtless Play: Using William Blake to Illuminate Authority and Agency Within *Grand Theft Auto: San Andreas*." *Games and Culture*. 2/2 (Apr 2007).

Salen, Katie and Zimmerman, Eric. *Rules of Play: Game Design Fundamentals*. Cambridge, MA: MIT Press, 2003.

Salter, Mark. "Geographical Imaginations of Video Games: *Diplomacy*, *Civilization*, *America's Army* and *Grand Theft Auto IV*." *Geopolitics*. 16 (2011).

Sample, Mark. "Virtual Torture: Videogames and the War on Terror." *Game Studies*. 8/2 (Dec 2008).

Samuel, Lawrence R. *The End of the Innocence: The 1964–1965 New York World's Fair*. New York: Syracuse University Press, 2010.

Sandomir, Richard. "'America's Game': The Real National Pastime." *New York Times*. 7 Nov 2004.

Schiesel, Seth. "Way Down Deep in the Wild, Wild West." *New York Times*. 16 May 2010.

Schlosser, Eric. "World War Three, By Mistake." *New Yorker*. 23 Dec 2016.

Schoon, Robert. "Tap That App: Hands-On With Trumpada—Latino Trump Backlash Comes to Android in a Game." *Latin Post*. 11 Sept 2015. https://www.latinpost.com/articles/78050/20150911/tap-that-app-hands-on-with-trumpada-latino-trump-backlash-in-android-game-form.htm.

Schulzke, Marcus. "Being a Terrorist: Video Game Simulations of the Other Side of the War on Terror." *Media, War and Conflict*. 6/3 (2013).

Schulzke, Marcus. "Moral Decision Making in *Fallout*." *Game Studies*. 9/2 (Nov 2009).

Schulzke, Marcus. "Virtual War on Terror: Counterterrorism Narratives in Videogames." *New Political Science*. 35/4 (2013).

"The Science and Poetry of the Light in Los Angeles." *New Yorker*. 9 Dec 2016. https://www.newyorker.com/culture/culture-desk/the-science-and-poetry-of-the-light-in-los-angeles.

Scimeca, Dennis. "*Grand Theft Auto* Hates America." *Salon*. 20 Sept 2013. https://www.salon.com/2013/09/20/grand_theft_auto_hates_america/.

Scott, Neil. "My Activist Second Life." *Red Pepper*. 28 Feb 2009. https://www.redpepper.org.uk/left-unity-in-second-life/.

Seabrook, John. "Game Master." *New Yorker*. 6 Nov 2006.

Sederstrom, Jotham. "Artist Axes 9/11 'Invaders' Work." *New York Daily News*. 24 Aug 2008.

Shaffer, Marguerite. "'The West Plays West': Western Tourism and the Landscape of Leisure." In William Deverell, ed., *A Companion to the American West*. Malden, MA: Blackwell, 2004.

Shapiro, Jerome. *Atomic Bomb Cinema: The Apocalyptic Imagination on Film*. New York: Routledge, 2001.

Shaw, Tony. *Hollywood's Cold War*. Edinburgh: Edinburgh University Press, 2007.

Shoop, Casey. "Corpse and Accomplice: Fredric Jameson, Raymond Chandler, and the Representation of History in California." *Cultural Critique*. 77 (Winter 2011).

Sloan, Kim. *A New World: England's First View of America*. London: British Museum, 2007.

Slotkin, Richard. *The Fatal Environment: The Myth of the Frontier in the Age of Industrialization, 1800–1890*. Middletown, CT: Wesleyan University Press, 1985.

Slotkin, Richard. *Gunfighter Nation: The Myth of the Frontier in Twentieth-Century America*. New York: Atheneum, 1992.

Slotkin, Richard. *Regeneration through Violence: The Mythology of the American Frontier, 1600–1860* Middletown, CT: Wesleyan University Press, 1973.

Smith, Dave. "Makers of the 9.11 Virtual Reality App Explain Why They Made the Controversial Game." *Business Insider* (UK). 29 Oct 2015.

Smith, Ed. "*Grand Theft Auto* and the Airbrushing of History." *Alphr*. 2 Feb 2016. http://www.alphr.com/games/1002581/grand-theft-auto-and-the-airbrushing-of-history.

Smith, Ed. "The Inglorious Nihilism of *Grand Theft Auto V*." *Kill Screen*. 19 May 2016. https://killscreen.com/articles/inglorious-nihilism-grand-theft-auto-v/.

Smith, Harrison. "Soviet Officer Headed off Nuclear War." *Washington Post*. 18 Sept 2017.

Smith, Henry Nash. *Virgin Land: The American West as Symbol and Myth*. Cambridge, MA: Harvard University Press, 1950.

Smith, Tony. "Star Trek: The Original Computer Game." *Register*. 3 May 2013. https://www.theregister.co.uk/2013/05/03/antique_code_show_star_trek/.

Smith, William. "Electronic Games Bringing a Different Way to Relax." *New York Times*. 25 Dec 1975.

Sneiderman, Phil. "Last Rites: City of Lost Souls Can't Escape Judgment Day." *Los Angeles Times*. 9 Feb 1991.

Snider, Mike. "Q&A: Mario Creator Shigeru Miyamoto." *USA Today*. 8 Nov 2010.

Soja, Edward. "Inside Exopolis: Scenes from Orange County." In Michael Sorkin, ed., *Variations on a Theme Park: The New American City and the End of Public Space*. New York: Hill & Wang, 1992.

Soja, Edward. *Postmodern Geographies: The Reassertion of Space in Critical Social Theory*. London: Verso, 1989.

Solnit, Rebecca. *River of Shadows: Eadweard Muybridge and the Technological Wild West*. New York: Penguin, 2003.

Sontag, Susan. *On Photography*. New York: Farrar, Straus and Giroux, 1977.

"Sony Delays *Syphon Filter 3*." *IGN*. 19 Sept 2001. http://uk.ign.com/articles/2001/09/19/sony-delays-syphon-filter-3.

Sorkin, Michael. *Exquisite Corpse: Writing on Buildings*. New York: Verso, 1991.

Spigel, Lynn. "Entertainment Wars: Television Culture after 9/11." In Dana Heller, ed., *The Selling of 9/11: How a National Tragedy Became a Commodity*. New York: Palgrave, 2005.

Stafford, Matthew. "Wargames to Screen." *Sonoma Index-Tribune*. 8 June 2017.

Stahl, Roger. "Have You Played the War on Terror?" *Critical Studies in Media Communication*. 23/2 (2006).

Steen, Francis, Patricia Greenfield, Mari Davies, and Brendesha Tynes. "What Went Wrong with *The Sims Online*: Cultural Learning and Barriers to Identification in a Massively Multiplayer Online Role-Playing Game." In Peter Vorderer and Bryant Jennings, eds., *Playing Video Games: Motives, Responses, and Consequences*. Mahwah, NJ: Erlbaum, 2006.

Steiner, Michael. "Frontierland as Tomorrowland: Walt Disney and the Architectural Packaging of the Mythic West." *Montana: The Magazine of Western History*. 48/1 (Spring 1998).

Stephens, Mark. *Three Mile Island*. New York: Random House, 1980.

"The Story Behind the Haunting 9/11 Photo of a Man Falling from the Twin Towers." *Time*. 8 Sept 2016.

Stuart, Keith. "*Grand Theft Auto 5*: Rockstar's Dan Houser on Los Santos and the future." *Guardian*. 17 Sept 2013.

Stuart, Keith. "*GTA5* Review: A Dazzling but Monstrous Parody of Modern Life" *Guardian*. 16 Sept 2013.

Stubblefield, Thomas. *9/11 and the Visual Culture of Disaster*. Bloomington: Indiana University Press, 2015.

Suellentrop, Chris. "Americana at Its Most Felonious." *New York Times*. 9 Nov 2012.

Swanson, Ana. "Why Amazing Video Games Could Be Causing a Big Problem for America." *Washington Post*. 23 Sept 2016.

Sweet, Sam. "Idling in Los Santos." *New Yorker*. 20 Sept 2013.

Tapley, Rebecca. *Designing Your Second Life*. Berkeley: New Riders, 2007.

"The Theme and Symbol of the Fair: The 140-Foot-High Unisphere." *New York Post*. 18 Nov 1963.

Thompson, Nicholas. "Inside the Apocalyptic Soviet Doomsday Machine." *Wired*. 21 Sept 2009.

Thoreau, Henry David. *Walden*. Chesterville: Kellscraft, 2010 [1854].

Thwaites, Reuben, ed. *Original Journals of the Lewis and Clark Expedition, 1804–1806*. 8 vols., New York, 1904.

Turner, Frederick Jackson. "The Significance of the Frontier in American History." [1893] In *The Frontier in American History*. 1920.

Udall, Stewart et al. "How the West Got Wild: American Media and Frontier Violence." *Western Historical Quarterly*. 31/3 (Autumn 2000).

Ulin, David. "There It Is. Take It." *Boom* (California). 3/3 (Fall 2013). https://boomcalifornia.com/2013/09/23/there-it-is-take-it/

Ulin, David, ed. *Writing Los Angeles: A Literary Anthology*. New York: Library of America, 2002.

"Uproar Grows over GTA Sex Scenes." *BBC News*, 26 July 2005. http://news.bbc.co.uk/1/hi/technology/4717139.stm.

Virilio, Paul. *Ground Zero*. London: Verso, 2002.

Voorhees, Gerald. "I Play, Therefore I Am: Sid Meier's *Civilization*, Turn-Based Strategy Games and Cogito." *Games and Culture*. 4/3 (2009).

Voorhees, Gerald, Joshua Call, and Katie Whitlock, eds. *Guns, Grenades and Grunts: First Person Shooter Games*. New York: Continuum, 2012.

Wainwright, Oliver. "New York's Twin Towers." *Guardian*. 20 May 2015.

Walker, Trey. "EA Offers Retailers Revised Red Alert 2 Box Art." *Gamespot*. 17 Sept 2001. https://www.gamespot.com/articles/ea-offers-retailers-revised -red-alert-2-box-art/1100-2812631/.

Ward, Trent. "*Blast Corps* Review." *Gamespot*. 20 Mar 1997. https://www .gamespot.com/reviews/blast-corps-review/1900-2544167/.

Watkins, Kevin. "Save a Little Water for Tomorrow." *New York Times*. 17 Mar 2006.

Weart, Spencer. *Nuclear Fear: A History of Images*. Cambridge, MA: Harvard University Press, 1988.

Welsh, Timothy. *Mixed Realism: Videogames and the Violence of Fiction*. Minneapolis: University of Minnesota Press, 2016.

Westfahl, Gary. *Space and Beyond: The Frontier Theme in Science Fiction*. Westport, CT: Greenwood, 2000.

Whalen, Zach. "Cruising in San Andreas: Ludic Space and Urban Aesthetics in *Grand Theft Auto*." In Nate Garrelts, ed., *The Meaning and Culture of Grand Theft Auto: Critical Essays*. Jefferson, NC: McFarland, 2006.

Whalen, Zach, and Laurie Taylor. *Playing the Past: History and Nostalgia in Video Games*. Nashville, TN: Vanderbilt University Press, 2008.

White, Josh. "It's a Video Game, and an Army Recruiter." *Washington Post*. 27 May 2005.

White, Richard. "Animals and Enterprise." In Clyne Milner, ed., *The Oxford History of the American West*. New York: Oxford University Press, 1994.

White, Richard, "Frederick Jackson Turner and Buffalo Bill." In James Grossman, ed. Essays by Richard White and Patricia Nelson Limerick. *The Frontier in American Culture: An Exhibition at the Newberry Library, August 26, 1994–January 7, 1995*. Berkeley: University of California Press, 1994.

White, Stewart Edward. *The Rules of the Game*. New York: Doubleday, 1910.

Wills, John. "Celluloid Chain Reactions: *The China Syndrome* and Three Mile Island." *European Journal of American Culture*. 25/2 (2006).

Wills, John. "Digital Dinosaurs and Artificial Life: Exploring the Culture of Nature in Computer and Video Games." *Cultural Values*. 6/4 (2002).

Wills, John. *Disney Culture*. New Brunswick, NJ: Rutgers University Press, 2017.

Wills, John. "Wagon Train to the Stars: Star Trek, the Frontier and America." In Doug Brode, ed., *Gene Roddenberry's Star Trek The Original Cast Adventures*. Lanham, MD: Rowman and Littlefield, 2013.

Wolf, Mark. *Before the Crash: Early Video Game History*. Detroit, MI: Wayne State University Press, 2012.

Wolfe, April. "Hey, White People: Michael Douglas Is the Villain, Not the Victim, in *Falling Down*." *LA Weekly*. 26 Apr 2017.

Wordsworth, Richard. "How *Deus Ex* Predicted the Future." *Kotaku*. 14 Apr 2017. https://kotaku.com/how-deus-ex-predicted-the-future-1616252703.

Yi, Matthew. "Online Game Bets on Self-Expression." *San Francisco Chronicle*. 23 June 2003.

Ziff, Lloyd. "I Love LA." *New York Times*. 29 June 2013.

Žižek, Slavoj. *Welcome to the Desert of the Real*. New York: Verso, 2002.

INDEX

Spielberg, Steven, 27, 59
Spigel, Lynn, 146
Spitting Image (TV series), 104
sports: in early video games, 2–3, 6–7, 30, 33–37; and *Tennis for Two*, 24–25; and *Wii Sports*, 208
Stagecoach (film), 64, 67
Stahl, Roger, 155–56
Stanley, Douglas Edric, 130
Star Trek (TV series), 54–55, 100, 168
Star Wars (film), 4, 56, 76, 229
States, 39
Statue of Liberty, 18, 38, 110, 118, 121–22, 143, 199–200, 218
Stephenson, Neal, 194
Stone, Oliver, 141
Storer, Jim, 53–54
Stowe, Harriet Beecher, 31
Strategic Defense Initiative (SDI), 88, 98
strategy games: board games, 30; *Civilization*, 19, 46, 49; Cold War themed, 101; God games, 204; *Oregon Trail*, 43; terrorist-themed, 148; Western-themed, 78
Street Fighter, 38, 208
Streets of Rage, 117
Stubblefield, Thomas, 112, 126, 128, 131, 134, 139
suburbia, 7, 32, 167, 193, 207, 234
Sum of All Fears, The (film), 146
Sunset Riders, 68–69
Super Columbine Massacre RPG!, 38, 224
Superman, 97
Super Meat Boy, 224
survival-horror games, 13, 43, 76, 110, 219–21
Suzuki, Yu, 40
Syphon Filter series, 121, 127
System Shock, 230

table-tennis, 2, 7, 26–27, 66. *See also* Pong
Taito, 1, 36, 55, 59
Tank, 58
Tapper, 5, 7, 33, 223
Taylor, Simon, 119
Tecmo Bowl, 34

Tekken 2, 117
television: and 9/11, 112, 115, 126, 132, 146–47; and early video games, 2–3, 5; and Magnavox, 1, 29; and recreation, 31, 33, 231; visual America, 11
Tempest, 55
tennis, 24–25, 30
Tennis for Two, 24–25, 28
Terminator, The (film), 100
terrorism, 18–19, 109–60, 209, 223
Terrorist, 148
Testament (film), 94, 98, 109
Texas Chainsaw Massacre (film), 97
text-based games, 34, 37, 46, 54–55, 95
Theatre Europe, 101
Them! (film), 28
theme parks: and *Grand Theft Auto*, 171–72, 173; links with video game industry, 5, 18, 30–31; and *Second Life*, 200; and virtual reality, 138
Theurer, Dave, 87–89, 93
Think (film), 21, 23
Thoreau, Henry David, 201–2
Three Mile Island (nuclear accident), 100, 124
Thunderbird Strike, 225
Tic-Tac-Toe, 24, 83
Time Crisis, 76
Tingler, The (film), 28
Tomb Raider, 76
Tom Clancy's Splinter Cell: Conviction, 156
Tom Clancy's The Division, 130–31
Tooth Protectors, 5
Towering Inferno, 142
Trip to the Moon, A (film), 28
Tron (film), 58, 96, 99, 230
Trump, Donald, 225–26
Trumpada, 225
Turner, Frederick Jackson, 52, 55, 229
Twin Peaks (TV series), 219
Twin Towers. *See* World Trade Center

Ultima Online, 193
United States Adventure, 38–39
UNIVAC (computer), 23, 25, 56, 86, 102–3